The Shipwright's Song

A Timeless Account of Shipbuilding Craftsmanship Through History

Callahan Fairbanks

Table of Contents

1 **Chapter 1: Ancient Mariners** **5**
1.1 Early Boats 7
1.2 Egyptian Shipbuilding 10

2 **Chapter 2: Classical Craftsmen** **13**
2.1 Greek Triremes 16
2.2 Roman Galleys 19
2.3 Naval Architecture 22

3 **Chapter 3: Viking Voyages** **27**
3.1 Longship Construction 29
3.2 Viking Shipyards 32
3.3 Maritime Expansion 35
3.4 Raid & Trade Vessels 38
3.5 Clinker Building 41

4 **Chapter 4: Medieval Marvels** **45**
4.1 Cog Ships 48
4.2 Carrack Design 51

5 **Chapter 5: Age of Exploration** **55**
5.1 Caravel Construction 58
5.2 Navigational Advancements 61

5.3 Global Trade Routes 64

6 Chapter 6: The Golden Age 69

6.1 Galleon Building 72

6.2 Shipyard Innovations 74

7 Chapter 7: Modern Methods 79

7.1 Steel Shipbuilding 81

7.2 Engineered Vessels 84

7.3 Modern Shipyards 87

7.4 Advanced Materials 90

7.5 Future of Shipbuilding 93

1 Chapter 1: Ancient Mariners

Early boats, rudimentary yet ingenious, represent humanity's initial foray into maritime exploration. Imagine the earliest vessels – simple rafts of reeds or logs lashed together, providing precarious passage across rivers and lakes. These weren't merely functional; they were a testament to human resourcefulness, a bridge between land and water, opening up new possibilities for trade, migration, and survival. The development of dugout canoes, carved from single tree trunks, marked a significant leap forward, offering greater stability and carrying capacity. These early craft laid the groundwork for future maritime advancements, shaping the very course of human history. Their simple construction, however, belied the profound impact they would have on civilization's expansion.

The ancient Egyptians, masters of engineering and organization, elevated boat building to an art form. Their sophisticated shipbuilding techniques, meticulously documented in tomb paintings and surviving fragments of vessels, reveal a remarkable understanding of naval architecture. They employed skilled artisans to construct impressive vessels, utilizing cedar wood imported from Lebanon, papyrus reeds for lighter craft, and advanced joinery methods to create watertight hulls. These boats, some reaching considerable lengths, were not only used for transporting goods along the Nile but also for maritime trade with neighboring civilizations. Their contributions extended beyond simple transport; the construction of large, ceremonial boats, often associated with pharaohs and deities, further highlights the significance they attributed to shipbuilding. The durability and craftsmanship of these boats are evidenced by the remarkable preservation of some ancient Egyptian vessels, providing invaluable insights into their techniques.

Egyptian shipbuilding transcended mere practicality; it was imbued with spiritual significance. The construction process, from the felling of trees to the final launch, was often imbued with ritual and ceremony, reinforcing the societal importance of these vessels. The skilled shipwrights, revered for their expertise, were an integral part of Egyptian society, contributing to the nation's prosperity and power through their exceptional craftsmanship. They played a crucial role in facilitating trade, connecting Egypt to distant lands and enabling the exchange of goods and ideas. Consider the logistical challenges: transporting massive amounts of lumber across vast distances, organizing large teams of workers, and meticulously crafting sturdy hulls capable of withstanding the rigors of river travel. It was a complex undertaking, requiring sophisticated planning, organization, and execution. The Nile, a lifeline of Egyptian civilization, became navigable thanks to their expertise. Their legacy extends far beyond their time, shaping subsequent naval technologies and leaving a lasting impression on the world of maritime history.

The transition from simple rafts to more complex structures like the Nile barges represents a pivotal moment in maritime technology. This evolution was not a sudden leap but rather a gradual process of experimentation, innovation, and refinement. Each improvement, however minor, contributed to increased seaworthiness, cargo capacity, and efficiency. Consider the introduction of sails, for instance, a transformative innovation that fundamentally altered seafaring. The adoption of new materials, tools, and techniques also played a crucial role in improving boat design and construction. The Egyptians, through their remarkable ingenuity, weren't merely constructing boats; they were building the foundations of a globalized world, connecting civilizations and cultures across vast distances. Their legacy continues to resonate today, influencing modern shipbuilding principles and techniques. The very essence of their achievements still provides inspiration for engineers and artisans.

1.1 Early Boats

The earliest boats weren't boats as we understand them. Imagine, instead of meticulously crafted hulls, simple rafts of reeds or logs lashed together. These rudimentary vessels, born from necessity, allowed humans to cross rivers, navigate lakes, and explore coastal waters. Their construction was straightforward; strength came from the sheer number of elements bound tightly together. Imagine the collective effort, the shared knowledge passed down through generations, culminating in a craft that offered a fragile bridge across watery expanses. These weren't voyages of exploration in our modern sense; they were survival, a constant dance between the currents and the will to survive.

The development of dugout canoes marked a significant advance. Hollowed-out tree trunks, a testament to both human ingenuity and physical strength, offered greater stability and carrying capacity. This simple act of carving a vessel from a single log transformed water travel. No longer were journeys limited to calm waters; dugouts could venture further, into more challenging environments. Consider the skill involved, the precise removal of wood to create a watertight vessel capable of supporting a person, their belongings, and perhaps even a small catch of fish. This was technology at its most fundamental, a direct interaction between humanity and nature.

Later, the use of animal hides stretched over wooden frames expanded the possibilities further. Imagine stretching taut, carefully prepared hides, creating a flexible yet surprisingly watertight hull. This simple technique, used for centuries, allowed for greater size and load-bearing capacity, and it opened new avenues for transportation and trade. The boats became more than just survival tools; they became vehicles for commerce, connecting communities and expanding horizons. This evolution wasn't a sudden leap; it was a gradual refinement, built on centuries of trial and error.

As civilizations developed, so did their shipbuilding techniques. The emergence of sewn-plank construction represented a significant leap forward. Individual planks, carefully shaped and joined together, created stronger, more durable vessels. This new approach allowed for larger, more complex hulls – capable of carrying heavier loads and braving harsher seas. Imagine the intricate process, the precise measurements, the craftsmanship required to fit each plank seamlessly into place. This wasn't just building a boat; it was creating a functional work of art.

The early use of sails provided yet another transformative development. Harnessing the power of the wind propelled boats to new destinations, opening up opportunities for exploration and trade on an unprecedented scale. Consider the rudimentary sails of early boats, perhaps simple squares of woven material, offering little control but providing a significant boost in speed and efficiency. This interaction with the natural elements, however basic, forever changed the dynamics of water travel. With a single change, a journey that may have taken days could now be completed in hours.

The shift from simple transportation to sophisticated maritime ventures was gradual. Early boat designs were dictated by the available materials and the immediate needs of the communities that built them. But each incremental improvement – the use of different materials, new construction techniques, even the introduction of sails – accumulated over time to create a fundamental shift in humanity's relationship with the sea. These early boats were more than just vessels; they were symbols of human ingenuity, testament to our ability to adapt, innovate, and conquer new frontiers. They were the foundation upon which all future maritime advancements were built.

The story of early boats is a story of continuous adaptation. We know now that boats weren't static forms – they changed and evolved constantly to suit the environments and needs of the people who crafted them. We can see this gradual refinement reflected in the archaeological record. From humble beginnings as simple rafts and dugouts, boats developed into more sophisticated and seaworthy craft. It's a testament to our relentless ingenuity, a journey that continues to this day.

1.2 Egyptian Shipbuilding

Ancient Egyptian shipbuilding, a testament to human ingenuity, relied heavily on readily available resources. The Nile River, the lifeblood of their civilization, provided abundant papyrus, a crucial material for early boats. These vessels, initially simple rafts, gradually evolved into more sophisticated designs, showcasing a remarkable understanding of buoyancy and stability. The construction involved intertwining papyrus reeds, often coated with resin or bitumen for waterproofing, creating buoyant and surprisingly durable crafts capable of navigating the river's currents and carrying significant cargo. This early ingenuity laid the foundation for later, more ambitious shipbuilding projects.

Later periods saw a shift towards wooden hulls, utilizing cedar imported from Lebanon or acacia grown domestically. The Egyptians mastered techniques of woodworking, shaping planks with tools like adzes and chisels. These planks, carefully joined using mortise and tenon joints or pegged together, formed the hull's framework. They demonstrated an understanding of hull curvature and structural integrity, building vessels with impressive size and load-bearing capacity. The use of internal framing, rudimentary but effective, enhanced the strength and rigidity of the larger ships. This development marked a considerable step forward in shipbuilding technology.

Beyond the hull, Egyptian shipbuilders paid meticulous attention to rigging and sails. Mast construction involved carefully selecting and shaping long, straight timbers, often strengthened with additional supports. Sails, usually made of woven linen or papyrus, were attached to yards, enabling the vessel to harness wind power for navigation. The Egyptians developed sophisticated rigging techniques, experimenting with different sail designs to maximize efficiency and maneuverability, especially important for navigating the river's variable winds and unpredictable currents. Their understanding of wind patterns was integral to effective river navigation.

The construction process often involved specialized labor, reflecting a sophisticated level of organization and division of labor within the shipbuilding industry. Evidence from archaeological finds, such as tomb paintings depicting shipbuilding activities and the discovery of well-preserved vessels, provides valuable insights into their methods. These depictions reveal skilled craftsmen working together, each with specific roles and responsibilities, ensuring a smooth and efficient production process. The collaborative nature of Egyptian shipbuilding speaks volumes about their social structure and the importance of skilled trades.

Interestingly, Egyptian shipbuilding wasn't limited to river navigation. Their vessels undertook voyages on the Red Sea and the Mediterranean, engaging in trade with neighboring civilizations. This maritime expansion necessitated the construction of larger, more seaworthy vessels, capable of withstanding the challenges of open water navigation. The design evolved to incorporate improved hull shapes, stronger structures, and more robust rigging, reflecting an adaptation to the changing demands of longer voyages. This seafaring prowess extended their influence and commercial reach.

The development of Egyptian shipbuilding reflects a continuous process of innovation and refinement. From simple papyrus rafts to sophisticated wooden vessels capable of long-distance voyages, their skills and knowledge steadily improved over centuries. Their contributions to shipbuilding laid the groundwork for future advancements, influencing the maritime traditions of neighboring cultures and shaping the development of seafaring across the Mediterranean world. This legacy endures, reflecting a remarkable story of human ingenuity and adaptation. The techniques and materials used reveal a high level of sophistication and resourcefulness. Their successes on the Nile and beyond are a testament to their mastery of the craft.

2 Chapter 2: Classical Craftsmen

The Mediterranean world, a cradle of civilization, also witnessed a remarkable flourishing of naval technology. Greek triremes, the epitome of ancient warships, dominated the seas. Their construction, a marvel of engineering for its time, involved the precise assembly of hundreds of precisely-hewn timber planks, fastened with bronze nails and pegged together. This meticulous craftsmanship, passed down through generations of skilled shipwrights, resulted in a vessel capable of incredible speed and maneuverability. These sleek, fast ships were essential to the rise of Greek city-states, fueling their trade and military ambitions across the Aegean Sea and beyond. The oarsmen, meticulously trained, were the very heart of the trireme's power; their synchronized movements, a testament to the discipline and coordination that were key to their military effectiveness.

Roman galleys, while less renowned for speed than their Greek counterparts, possessed a different kind of power. Their larger size and robust construction made them formidable in battle, ideal for transporting legions and supplies across the vast expanse of the Roman Empire. The Romans refined shipbuilding techniques, developing more efficient methods of construction, and expanding the use of various materials to create larger and sturdier ships. Their influence stretched far beyond military applications, influencing the design and construction of merchant vessels that fueled the economic engine of the empire. This maritime dominance facilitated the expansion of trade, the rapid movement of troops, and ultimately, the consolidation of Roman power throughout the Mediterranean.

Naval architecture, in its nascent stages, was crucial to both Greek and Roman maritime success. While lacking the sophisticated mathematical modeling of later eras, shipwrights of the classical world possessed an intuitive understanding of hydrodynamics and structural integrity. They meticulously analyzed the relationship between hull shape, stability, and speed, their knowledge honed through generations of trial and error, constantly refining their methods and designs. This empirical approach, combined with an innate understanding of materials and their properties, allowed them to build vessels capable of withstanding the rigors of the sea, making voyages across the Mediterranean Sea routine. This period laid the foundation for future advancements in shipbuilding.

The skill of the classical shipwright extended beyond simply constructing vessels. They were masters of woodworking, metalworking, and rigging. Their knowledge encompassed a wide spectrum of crafts, necessitating an unparalleled degree of expertise. The construction of a single ship was a complex process, involving specialized skills and intricate coordination. The shipwright had to be a designer, engineer, project manager, and craftsman all rolled into one. This multifaceted role demanded a unique blend of theoretical knowledge and practical skill, fostering a deep respect for their craft, which in turn was crucial to creating reliable, efficient and powerful warships and merchant vessels. This skill, along with their understanding of materials, and their innovation in design, truly shaped the destinies of the ancient world's empires. A significant aspect was not only about the functionality but also the aesthetics of ship building; the beauty and design, the pride of their craftsmanship was reflected in the very appearance of the vessel, a testament to their skill and the pride they took in their work. The shipwrights' expertise helped to propel the development and the spread of both culture and civilization.

The classical period's legacy in shipbuilding is profound. Their innovations, although seemingly primitive by modern standards, established crucial principles of naval architecture that would shape shipbuilding practices for centuries to come. These craftsmen weren't just building ships; they were forging the very foundations of a global maritime network that would drive economic development and power shifts throughout history. Their practical knowledge, honed over generations, remains a testament to the power of human ingenuity and the enduring impact of skilled craftsmanship. The study of their methods provides valuable insights into the fundamental principles that continue to inform the field of naval architecture. We must understand this heritage not merely as a historical curiosity but also as an invaluable source of inspiration and knowledge for aspiring shipwrights in every generation. They built, with their bare hands and the tools of their age, the vessels that changed the course of history, providing a crucial step forward for humanity in its exploration of our oceans. The world learned to navigate it through these men's hands.

2.1 Greek Triremes

The Greek trireme, a marvel of ancient shipbuilding, dominated the Mediterranean for centuries. Its effectiveness stemmed not just from its speed and maneuverability, but also from sophisticated construction techniques, reflecting a deep understanding of naval architecture. Unlike earlier vessels relying on simple frameworks, the trireme employed a complex system of precisely fitted planks, creating a remarkably strong and lightweight hull. This advanced construction allowed for the incorporation of its signature three banks of oars, a feature that significantly enhanced its speed and power.

Its design, honed over generations, involved a carefully planned arrangement of timbers. The keel, a robust spine running the length of the vessel, provided structural integrity. Strong ribs, known as frames, were meticulously attached to the keel, providing support and shaping the hull's form. The outer planking, carefully selected for its strength and flexibility, was fitted tightly together to ensure watertightness, a crucial factor in a ship that spent considerable time at sea. The hull's design, sharply pointed at both ends, minimized water resistance, optimizing its speed and maneuverability, critical assets in naval warfare.

Construction involved specialized skills and precise measurements. Experienced shipwrights, masters of their craft, meticulously shaped each timber using hand tools. The process of joining these planks required expertise; skilled craftsmen fitted them together with such precision that minimal caulking was needed, another testament to their mastery. This emphasis on meticulous craftsmanship was essential; the success of the entire vessel rested on the precision of these seemingly minor details. Each individual plank was critical, not only to the structural integrity of the ship but also to its overall efficiency and performance.

Beyond the hull, the trireme's oar system was a masterpiece of engineering. Three levels of oarsmen, each row working in perfect synchrony, propelled the vessel at remarkable speeds. The oars themselves, long and strong, were carefully designed to minimize strain and maximize power. The coordination between the oarsmen, achieved through rigorous training and disciplined rowing, played a pivotal role in the vessel's overall efficacy. The successful deployment of a trireme depended heavily on effective management, training and organization of the rowing crew. The speed and power of the trireme was not merely a product of the craftsmanship involved in the ship's construction, but also a reflection of the efficient management of its human capital.

The trireme's impact extended far beyond its military applications. Its design and construction techniques influenced subsequent shipbuilding practices throughout the Mediterranean world. The speed and efficiency of the trireme contributed significantly to the expansion of trade, as its quick transit times shortened the time required to transport goods and people, thereby fostering economic growth and contributing to the cultural exchange that characterized the era. The evolution of shipbuilding techniques is inextricably linked to the growth of commerce and cultural interaction across the Mediterranean basin. The Greek trireme, a testament to the ingenuity of ancient shipwrights, left a lasting legacy on both the military and commercial domains of the ancient world.

The study of surviving trireme remains, albeit fragmented, provides invaluable insight into the methods and materials employed. Archaeologists painstakingly reconstruct these fragments, piecing together the clues to understand the shipbuilding methods. The sheer amount of work required highlights the dedication and commitment of those involved in uncovering secrets of ancient ship technology. This work not only deepens our understanding of the trireme but also reveals the remarkable craftsmanship of ancient shipwrights. Every piece of evidence, however small, yields significant insights into the complexity of the ancient shipbuilding process, enhancing our comprehension of maritime history and naval technology.

Moreover, analyzing depictions of triremes on ancient pottery and sculptures offers additional clues about their construction and appearance. These artistic representations, while not always perfectly accurate, provide valuable contextual information and help us visualise the trireme in its historical setting. These artistic interpretations offer a unique perspective and deepen our understanding of the cultural significance of the vessel. By studying the various depictions, we can start to understand the role this powerful war machine played in the lives of the Ancient Greeks.

In conclusion, the Greek trireme represents a high point of ancient shipbuilding. Its speed, maneuverability, and strength resulted from meticulous design and skilled craftsmanship. Its impact extended far beyond military applications, influencing trade, travel, and communication across the Mediterranean. The study of existing remains and historical records continues to illuminate the ingenuity and technological prowess of the ancient shipwrights who created this iconic vessel. The legacy of the trireme endures, offering a captivating glimpse into the ingenuity and technological capabilities of a civilization that fundamentally shaped the course of history.

2.2 Roman Galleys

Roman galleys, the backbone of the Roman navy, weren't merely boats; they were sophisticated warships, meticulously crafted for speed, maneuverability, and devastating power. Their design, a testament to Roman engineering prowess, significantly influenced naval warfare in the Mediterranean for centuries. Unlike their Greek predecessors, Roman galleys often featured a larger, broader hull, offering increased cargo capacity and stability, vital for supplying legions across vast distances.

The construction of these vessels involved a complex process, beginning with the selection of high-quality timber. Pine, cedar, and oak, renowned for their strength and durability, were carefully chosen and shaped. Roman shipwrights, skilled artisans, utilized advanced tools and techniques, including mortise and tenon joinery, to assemble the hull planks. This precise method ensured a watertight and robust structure capable of withstanding the rigors of seafaring. The hulls were typically strengthened with internal frames and reinforced with metal brackets where necessary.

Rowing power formed the heart of these impressive machines. Unlike the relatively shallow-draft Greek triremes, Roman galleys, particularly the larger types, employed multiple banks of oars, allowing for greater speed and thrust. Each oar was manned by a skilled rower, with a complex system of coordination ensuring synchronized movement for optimal efficiency. These massive vessels required hundreds of oarsmen, a testament to the Roman ability to organize and direct large-scale operations. The sheer scale of manpower needed highlights their importance in the Roman military and economic machine.

The galleys weren't merely vessels of war, they served crucial economic roles. Their significant cargo capacity facilitated the transportation of vital resources, goods, and troops across the Roman Empire's extensive territories. This vital trade network, supported by the galleys, played a critical role in maintaining the empire's stability and prosperity. From grain shipments to military supplies, these vessels were instrumental in sustaining daily life throughout the empire's far-reaching provinces.

Beyond their roles in transport and warfare, Roman galleys were symbols of power and prestige. Their size and impressive construction reflected the might of the Roman state. Their presence in ports across the Mediterranean served as a constant reminder of Rome's dominance, ensuring that the image of Rome's power would never be far from sight. The visual impact of these majestic vessels is a significant aspect that can't be underestimated, and their contribution to Roman political and social life is important to understand.

However, constructing these impressive ships came with its own difficulties. The sheer scale of the project required significant resources, both material and human. Vast quantities of timber, metal, and other raw materials needed to be sourced, transported, and processed. Moreover, the workforce needed to construct and maintain these vessels was immense, calling for skilled shipwrights, oarsmen, and support personnel. This labor-intensive enterprise presented constant logistic challenges; everything had to function perfectly to ensure the timely completion of these crucial naval assets.

Moreover, the maintenance of these vessels presented significant ongoing challenges. Regular repairs were essential to maintain their seaworthiness. Roman shipyards employed a large number of skilled workers dedicated to this essential task, constantly patching up damage caused by the wear and tear of prolonged use at sea and harsh environmental conditions.

It is important to mention that the vastness of the Roman navy demanded a significant investment of resources. The Roman economy heavily relied on its naval prowess; maintaining this substantial fleet required an intricate logistics network and extensive financial resources.

The impact of Roman galleys extended far beyond the Mediterranean. Their design influenced subsequent naval architecture, inspiring shipbuilders for centuries to come. Features such as their hull design, rowing systems, and the overall size and complexity directly influenced the design of later warships. Their impact on naval history, as well as wider maritime history, is significant and deserves in-depth study.

In conclusion, Roman galleys represent a significant achievement in ancient shipbuilding. Their construction, operation, and overall impact demonstrate Roman engineering prowess and organizational capabilities. From their use in warfare to their role in trade and their contribution to the empire's political image, their significance remains undeniable. Studying their construction reveals insights into Roman society, technology, and the intricate workings of a powerful empire. The legacy of Roman galleys continues to fascinate historians, engineers, and maritime enthusiasts alike, and future research will undoubtedly continue to shed light on their historical importance. Their story illustrates the importance of naval power in the ancient world and the complexity of the engineering needed to achieve naval dominance.

2.3 Naval Architecture

Understanding the principles of naval architecture is crucial for appreciating the evolution of shipbuilding. It's not merely about constructing a vessel; it's about marrying form and function to create a structure capable of withstanding the immense forces of the sea. Consider the challenges: buoyancy, stability, structural integrity – all must be meticulously addressed to ensure a seaworthy craft. This discipline draws upon physics, mathematics, and engineering, requiring a deep understanding of hydrostatics, hydrodynamics, and material science. Master shipwrights, throughout history, intuitively grasped these concepts, their knowledge passed down through generations. Today, naval architects employ sophisticated computer-aided design (CAD) and computational fluid dynamics (CFD) to achieve similar feats, though the fundamental principles remain unchanged.

Early shipbuilders relied heavily on empirical knowledge. Observations of natural forms, from the buoyant properties of logs to the streamlined shapes of fish, informed their designs. Trial and error played a significant role, with failures providing valuable lessons learned that shaped future constructions. The Egyptians, for instance, mastered the art of building large vessels using reeds and papyrus, demonstrating an intuitive grasp of buoyancy and structural stability. Their vessels, while seemingly simple, were remarkably effective for navigating the Nile and engaging in trade across the Mediterranean. Their success underscores the importance of practical experimentation in the development of naval architecture.

The Greeks and Romans significantly advanced the science. The development of the trireme, a warship propelled by oarsmen in three tiers, stands as a testament to their sophistication. This wasn't simply a matter of adding more oars; careful consideration was given to the overall hull shape, the placement of oars and rowers, and the structural integrity required to withstand the forces generated by hundreds of rowers working in unison. The Romans, known for their practicality, further refined these designs, adapting them for both naval combat and transport. Their galleys, while differing from Greek triremes in detail, exemplified the continuing evolution of naval architecture principles toward greater efficiency and seaworthiness.

The advent of the medieval period brought about new challenges and innovative solutions. The cog, a large cargo vessel, demanded improvements in hull design to accommodate larger cargo loads and withstand the stresses of longer voyages. Carracks, larger and more seaworthy, represented a further evolution. These ships required more sophisticated structural elements to maintain their stability in the face of heavy seas. Improvements in rigging, mast placement, and sail design were also key developments in the medieval era, highlighting the interdisciplinary nature of naval architecture. It's a testament to the ingenuity of medieval shipwrights that they were able to construct vessels capable of extensive voyages across the oceans.

The Age of Exploration witnessed a dramatic shift in shipbuilding. The caravel, known for its maneuverability and seaworthiness, played a pivotal role in expanding European trade and exploration. This relatively small vessel was designed for speed and efficiency, reflecting a more nuanced understanding of hydrodynamics. Its design incorporated innovative features, including lateen sails which allowed for sailing close to the wind – a substantial advantage in navigating unpredictable ocean currents. Such developments highlight the interplay between design, functionality, and material selection, underscoring the dynamic nature of naval architecture.

The Golden Age of Sail saw the construction of impressive galleons, majestic warships that dominated the seas. Galleons were larger and more heavily armed than their predecessors, requiring careful consideration of hull design, weight distribution, and structural integrity to ensure stability under the weight of cannons and sails. The use of advanced construction techniques and materials resulted in larger, more durable, and more capable vessels. Shipbuilders in this era meticulously understood the delicate balance between speed, firepower, and stability, demonstrating a high degree of sophistication in naval architecture.

Modern naval architecture utilizes advanced computational tools to analyze vessel designs. Engineers now utilize computer-aided design (CAD) software to create detailed three-dimensional models of ships, simulating their performance under various conditions. Computational fluid dynamics (CFD) software allows for the analysis of water flow around the hull, optimizing the design for speed and efficiency. These advancements provide a significant advantage over the empirical approaches of earlier eras, enabling the creation of highly efficient and seaworthy vessels. The integration of these modern tools hasn't changed the underlying principles; instead, it's allowed for a far more precise understanding and execution of those fundamentals.

The shift towards steel construction brought a revolution in shipbuilding. Steel offered superior strength and durability compared to wood, enabling the construction of larger and more resilient vessels. Moreover, the introduction of mass production techniques streamlined the construction process, greatly reducing costs. This is crucial because it represents a transition from artisanal craft to mass production, which has also influenced other areas like manufacturing and assembly lines, showing again the interaction between seemingly unconnected fields. However, the underlying principles of naval architecture still apply; the optimal design must still be one that maximizes stability, buoyancy and minimizes drag to ensure the structure's safety and efficiency.

The future of naval architecture is focused on sustainability and efficiency. The rising costs of fossil fuels are pushing the industry to explore alternative energy sources, such as wind and solar power. Additionally, there's a growing emphasis on reducing the environmental impact of shipping. Innovations in materials science may lead to the use of lighter, stronger, and more sustainable composite materials. These advancements will not only increase the efficiency of ships but also reduce their carbon footprint, demonstrating that naval architecture is constantly evolving to meet the challenges and demands of the modern world. The underlying principles of naval architecture will continue to be crucial for future designs, but they will be applied with a greater focus on sustainability and efficiency.

3 Chapter 3: Viking Voyages

The Viking longship, a marvel of maritime engineering, wasn't merely a vessel; it was an extension of Viking life, a tool for conquest, exploration, and trade. Its slender, elegant form, deceptively strong, belied the sophisticated craftsmanship that went into its creation. The distinctive clinker-built hull, with overlapping planks fastened together with rivets and wooden pegs, provided both flexibility and strength, allowing the ships to navigate shallow rivers and withstand the pounding of open ocean waves. This wasn't simply a matter of joining planks; it demanded precision, skill, and an intimate understanding of wood's properties. The selection of timber, predominantly oak for its durability and resistance to decay, was critical; each plank was carefully shaped and fitted, a testament to the shipwright's artistry. The mastery of this technique represented generations of accumulated knowledge passed down through apprenticeships and family traditions.

Construction wasn't confined to a single location. Viking shipyards, often situated near suitable timber resources and waterways, were bustling centers of activity. These weren't standardized factories; rather, they were dynamic spaces, reflecting the needs and resources of each community.

The scale varied significantly, ranging from small workshops serving local needs to larger sites capable of producing multiple vessels simultaneously. Archaeological evidence suggests highly organized production, with skilled specialists handling different stages of the construction process, each contributing to the final vessel's robustness and seaworthiness. These shipyards weren't just about building ships; they fostered community and represented a significant aspect of Viking social and economic life. The longships' design facilitated extraordinary maritime expansion. Their shallow draft enabled exploration of inland waterways and coastal regions, giving the Vikings access to previously unreachable areas. Their speed and maneuverability, coupled with their seaworthiness, allowed them to conduct long-range voyages across vast stretches of ocean, reaching from the coasts of North America to the shores of the Caspian Sea. These weren't mere navigational feats; they were crucial to the Vikings' ability to trade, raid, and settle new territories, reshaping the political and cultural landscapes of Europe and beyond. The impact of the longships was far-reaching and enduring, significantly shaping the course of history.

However, Viking longships weren't uniformly built. While some were specifically designed for warfare, boasting a higher number of oars and a platform for warriors, others were built primarily for trade. These vessels prioritized cargo capacity over speed, and their construction often reflected that different purpose, with larger holds and fewer oars. The differences in design demonstrated the Vikings' adaptability, their ability to adjust their shipbuilding techniques to meet varied needs. It was not a simple "one size fits all" approach; rather a nuanced and sophisticated understanding of how design choice dictated function and effectiveness.

The clinker building technique, while complex, provided distinct advantages. The overlapping planks allowed for flexibility in the hull, enabling the ship to flex under the stress of waves without breaking. This inherent resilience, combined with the strength of the oak timbers, made these ships exceptionally seaworthy, capable of withstanding the rigors of long and demanding voyages. The method also made repairs relatively straightforward; replacing a damaged plank was less disruptive than in carvel-built ships where planks were edge-to-edge. This practicality had a significant impact on the long-range journeys these ships undertook, allowing for effective repairs at sea or in makeshift ports. The enduring legacy of the clinker method continues to influence boat building even today. The careful fitting of each plank, the precise placement of each rivet, each element played a critical role in creating a functional and enduring vessel. The enduring design of these ships is a testament to the Vikings' understanding of naval architecture, showcasing the ingenuity and skill of their shipwrights. The Viking voyages, powered by these expertly crafted longships, left an indelible mark on history, demonstrating the integral role shipbuilding played in their successes.

3.1 Longship Construction

The construction of a Viking longship was a testament to skilled craftsmanship and meticulous planning. Unlike later vessels built using frame-first techniques, the longships were created using clinker building, a method where overlapping planks were fastened together to form the hull.

This process, demanding precision and experience, began with the selection of carefully chosen timber, typically oak for its strength and durability. Each plank was shaped with adzes and knives, ensuring a snug fit against its neighbor. The planks' carefully crafted edges, thinned at the overlaps (called 'clinker'), created the hull's distinctive appearance. This technique, while seemingly simple, allowed for a flexible yet robust hull, ideal for navigating the often turbulent waters of the North Sea and beyond. The shipwrights' knowledge of wood properties was paramount. The choice of wood, its grain, and its moisture content were all critical factors influencing the ship's final strength and seaworthiness. The master shipwright's expertise guided the entire process.

Fastening these planks was a complex task, requiring specialized tools and considerable skill. Iron rivets, hammered through pre-drilled holes, secured each plank to the adjacent one, creating a watertight seal. The process required a coordinated team effort, with workers meticulously positioning and securing each plank to ensure the hull's overall integrity. The ship's stem and stern posts, meticulously carved from a single piece of wood, provided the crucial structural support for the hull. These posts formed the skeleton around which the ship was built, acting as anchors for the planks. Their careful placement dictated the ship's overall shape and performance characteristics. The keel, often a single massive piece of wood, added significant structural strength to the vessel.

The design itself was ingenious. The shallow draft enabled the longships to navigate shallow rivers and coastal waters, granting them access to inland areas inaccessible to deeper-draft vessels. Their narrow beam, combined with a relatively high length-to-beam ratio, provided speed and maneuverability, crucial attributes for both raiding and trading. The high sides, while offering protection from the elements, weren't excessive, maintaining a balance between protection and speed. The longships' construction involved more than just the hull. The construction of the deck, ribs, and planking demanded equal precision. The ribs, curved wooden supports, provided the frame for the hull's shape. These are fitted inside the hull, between the planking. It is important that there's a spacing of a few centimeters between the ribs to allow the ship to flex slightly without breaking. The deck, typically made of wooden planks, provided a platform for crew and cargo. Careful consideration went into the placement of oar ports, the openings through which oars were passed, to ensure efficient rowing. The mast and sail provided additional propulsion, allowing longships to take advantage of favorable winds. The use of sails expanded the operational range and capabilities of the longship, allowing for greater distances to be covered during voyages.

Building a longship was a communal effort. Shipyards, often located near suitable timber sources and waterways, became centers of activity. Skilled shipwrights, carpenters, and other laborers worked together, sharing knowledge and skills passed down through generations. The construction process involved careful planning, precise measurements, and coordinated teamwork. Master shipwrights, possessing years of experience, oversaw every aspect of the construction, ensuring quality and efficiency. Their expertise wasn't solely limited to technical skills but also encompassed knowledge of materials, design, and maritime conditions. Their guidance was vital in ensuring the construction of a seaworthy vessel. The time taken to build a longship varied, influenced by factors such as size and the availability of materials and labor. The knowledge was not codified. It was passed down from one generation to the next, through apprenticeships and hands-on experience.

The successful completion of a longship represented not only the culmination of skilled labor but also the embodiment of Viking maritime prowess. These vessels enabled the Vikings to explore, trade, and raid across vast distances, making their impact felt across Europe and beyond. The construction techniques developed, the materials selected, and the overall design all reflected a deep understanding of shipbuilding principles and maritime environment. These weren't just vessels of transportation; they were instruments of power and exploration. They played a pivotal role in Viking society, facilitating their maritime expansion, trade networks, and exploration, and shaping the history of northern Europe. Studying longship construction, therefore, offers a compelling insight into Viking culture, technology, and adaptation to their environment. The impressive feats of seamanship and exploration accomplished by the Vikings were significantly influenced by their mastery of ship construction.

3.2 Viking Shipyards

Viking shipyards weren't simply places where ships were built; they were vibrant hubs of activity, microcosms of Viking society reflecting their culture, technology, and economy. Located near suitable timber resources and waterways, these yards were often situated in sheltered bays or inlets, providing protection from the harsh North Sea weather. Skilled craftsmen, shipwrights, and laborers worked together, their expertise passed down through generations, a testament to the importance of maritime traditions within Viking communities. The rhythmic clang of hammers against wood, the rasp of adzes shaping planks, and the shouts of coordination formed the soundtrack of these crucial locations. They were, essentially, the engine room of Viking expansion.

The construction process itself was a testament to Viking ingenuity. Unlike later shipbuilding techniques that employed advanced tools and precise measurements, Viking shipwrights relied on a combination of skilled craftsmanship and intuitive understanding of wood's properties. They favored the durable oak, carefully selecting and felling trees that would provide the ideal strength and flexibility for their longships. The clinker building technique, characterized by overlapping planks fastened together with rivets, was a hallmark of Viking shipbuilding. This flexible construction allowed their ships to withstand the stresses of navigating rough seas and shallow waters, contributing significantly to their success in exploration and trade. The process started with the creation of sturdy frames, then the meticulous fitting of planks, secured by carefully positioned rivets. It was labor-intensive, demanding precision and a deep understanding of both material science and naval architecture.

The size and complexity of Viking shipyards varied depending on the scale of the community and its maritime activities. Smaller settlements might have had a modest yard capable of building and repairing only a few ships annually. Larger settlements, or those with a strong naval tradition, possessed extensive shipyards capable of producing multiple ships concurrently. These larger facilities required a highly organized workforce, potentially incorporating specialists for various tasks like plank-making, riveting, and rigging. Evidence suggests that shipyards were often near other vital infrastructure—forges, wood-processing areas, and storage spaces for materials and tools.

The economic impact of Viking shipyards extended beyond the immediate construction process. They were centers of trade, drawing resources from far-reaching networks. Oak timber, iron for rivets, tar for caulking, and rope for rigging all contributed to a complex system of economic exchange and interconnectedness. Shipyards fostered skilled employment, supporting families and contributing to the overall prosperity of Viking communities. Moreover, the longships themselves represented a significant investment of resources, both human and material. Their production and use directly influenced trade routes, raiding expeditions, and ultimately, the overall socio-political landscape of the Viking age. These ships weren't just vessels; they were instruments of exploration, trade, and power, with their creation at the heart of the Viking's world.

Furthermore, the design of Viking longships reveals the sophisticated understanding of hydrodynamics possessed by Viking shipwrights. Their vessels were remarkably efficient, combining speed with remarkable stability and seaworthiness. Longships, with their shallow drafts, could navigate rivers and coastal waters, while their streamlined hulls allowed for efficient travel over open seas. The placement of the mast, the shape of the sail, and the design of the rudder all played crucial roles in optimizing performance. This combination of features enabled Vikings to explore and raid across vast distances, making them a formidable presence in the waters of Europe and beyond. The design wasn't just about aesthetics; it was a testament to practical knowledge and effective problem-solving. It also suggests the shipyards were not just sites of production but also of ongoing innovation and improvement to designs.

Beyond their immediate contribution to Viking expansion, the shipyards offer valuable insight into the social structure of Viking society. The organization and management of such complex endeavors suggest a degree of social stratification and specialized labor. Evidence of specialized tools, standardized construction techniques, and the scale of some yards points to the existence of skilled professionals, project managers, and perhaps even a division of labor based on expertise. The shipyards were, therefore, a reflection of not only the technological capabilities of the Vikings but also the organization and social dynamics within their communities. They were a testament to their social structures. Studying them illuminates how their society was built and their place in the larger context of history.

3.3 Maritime Expansion

Viking maritime expansion wasn't solely about raiding; it was a complex interplay of exploration, trade, and conquest, all facilitated by their remarkable shipbuilding prowess. Their longships, slender and elegant yet remarkably seaworthy, were masterpieces of engineering, perfectly adapted to the shallow waters and unpredictable currents of the North Sea and beyond. The clinker-built construction, overlapping planks secured with rivets, allowed for a flexible hull capable of navigating both open ocean and narrow rivers, a versatility crucial for their multifaceted ambitions.

The construction of these vessels was far from a solitary endeavor. Viking shipyards, though often less formally structured than later shipyards, were bustling hubs of activity, bringing together skilled craftsmen—shipwrights, carpenters, blacksmiths, and rope makers—who collaborated to produce these impressive seafaring machines. The social structure within these shipyards likely reflected the broader Viking society, a hierarchical system where experience and skill determined status and responsibility. We can imagine the master shipwright overseeing the work, his expertise ensuring the quality and seaworthiness of each vessel.

Beyond the technical aspects, Viking maritime expansion was also a product of their sophisticated navigational skills. While the details of their navigational techniques remain somewhat obscure, their ability to navigate vast distances across the open ocean, relying on celestial observation and the knowledge of sea currents and winds, is a testament to their seamanship. This ability, combined with the speed and maneuverability of their longships, enabled them to raid distant coastal settlements, establish trading posts, and explore uncharted territories. Their voyages weren't merely acts of aggression; they were acts of exploration that fundamentally reshaped the political and cultural landscape of Europe.

The impact of Viking maritime expansion extended far beyond the immediate reaches of their raids and settlements. The exchange of goods and ideas along their trade routes fostered cultural exchange and technological diffusion. Viking longships carried not just warriors and plunder, but also merchants who traded furs, timber, slaves, and other goods throughout Europe and beyond. These trade connections strengthened economic ties between different regions and contributed to the formation of larger, more interconnected networks. The spread of Norse culture and language, evident in place names and loanwords throughout the North Atlantic region, is another important legacy of this period.

It wasn't just about conquest and trade though; it also involved intricate political strategies. Viking raids were often carefully planned and targeted, aimed at strategically weakening or undermining rival powers. The element of surprise and the speed and maneuverability of their ships were critical factors in their success. Yet, alongside their military campaigns, Vikings also established settlements and trading posts, demonstrating a capacity for both brutal aggression and long-term strategic planning. Their success illustrates the significant role that maritime capabilities played in shaping the political landscape of early medieval Europe. The ability to project power across vast distances, combined with adaptable ship design and skilled seamanship, allowed the Vikings to establish a far-flung network of influence and trade that extended from North America to the Caspian Sea. The legacy of their maritime expansion is deeply woven into the fabric of European history.

The creation of a seafaring culture amongst the Vikings also played a crucial part in their expansion, promoting collaboration and community. The construction and maintenance of longships required a communal effort, fostering the development of complex social structures and a shared understanding of the sea. This, combined with their advanced navigational knowledge and resilient ship construction, allowed the Vikings to push the boundaries of exploration, leaving an undeniable mark on the history of early medieval Europe. This wasn't a one-off occurrence, either; it was a systematic process that unfolded over decades, demonstrating the Vikings' ability to combine sophisticated shipbuilding with the adaptability to exploit various political and economic factors. The combination of seafaring culture, advanced shipbuilding techniques, and tactical flexibility ultimately shaped the Vikings' expansion into a major power across the North Atlantic and beyond. Their success serves as a powerful example of how maritime capabilities can influence the course of history.

3.4 Raid & Trade Vessels

Viking longships weren't solely instruments of war; they were also remarkably adaptable trade vessels. Their design, a testament to Viking ingenuity, facilitated both swift raids and lucrative commercial voyages. The shallow draft, a key characteristic, allowed them to navigate rivers and coastal waters, accessing inland trade routes inaccessible to deeper-hulled ships. This versatility is a crucial aspect often overlooked in discussions focusing solely on their military prowess. The speed and maneuverability, essential for raiding parties, also proved advantageous in the competitive world of maritime trade, enabling faster delivery of goods and quicker returns on investment.

Consider the construction materials. While oak was favoured for its strength and durability, the Vikings cleverly utilized readily available resources like pine and fir for specific parts of the vessel. This pragmatic approach speaks volumes about their resourcefulness and economic understanding. This wasn't simply about raiding and plunder; it was a sophisticated strategy encompassing both economic and military dominance. Their adaptability extended to cargo capacity. While capable of carrying substantial war supplies, they could be easily adapted to transport furs, timber, slaves and other valuable goods depending on the voyage's purpose.

The ship's design further supported its dual role. The characteristic clinker-built hull, with its overlapping planks, provided structural strength and flexibility, crucial for navigating rough seas and shallow waters. This was an adaptation that increased durability in often harsh conditions. The robust construction not only withstood the battering of sea travel but also provided a stable platform for both military maneuvers and the safe transport of trade goods. The relatively narrow beam, while reducing cargo capacity compared to later vessels, contributed significantly to the ship's speed, important for both offensive and commercial operations.

The sheer range of the Viking longships is astonishing. Their ability to cross vast stretches of ocean, from the Baltic Sea to the Mediterranean, to North America's shores, stands as a monument to their navigational skills and the quality of their shipbuilding. Such remarkable voyages would have been impossible without meticulously crafted vessels designed for both endurance and speed. They were instruments of exploration as much as of warfare and trade, extending the Vikings' reach across vast distances, opening opportunities for both expansion and exchange.

The economic impact of these vessels on Viking society was significant. The success of raiding expeditions provided wealth and resources, fueling further shipbuilding and expanding trading opportunities. A well-functioning trading network required dependable and capable vessels, and the longships were ideally suited for this purpose. The cycle of raiding and trading generated wealth that fuelled further expansion, contributing to the growth and development of Viking settlements and kingdoms across Europe and beyond. The economic success of the Vikings was inextricably linked to their shipbuilding prowess and the adaptability of their longships.

The longships, therefore, represent a fascinating case study in the interaction between military might and economic prosperity. They were not merely tools of war but also vital components of a thriving trading network. Their versatility underscores the deep understanding the Vikings possessed of naval architecture, not just for combat, but also for creating a sustainable and successful society. The integration of military and economic functions within a single vessel highlights the sophisticated nature of Viking maritime culture. They were practical, adaptive, and effective, leaving behind a legacy of naval innovation that echoes through maritime history. This intricate relationship between raiding and trade showcases the nuanced strategy and exceptional craftsmanship that underlay Viking success.

3.5 Clinker Building

Clinker building, a technique central to Viking shipbuilding, represents a sophisticated approach to wooden hull construction. Unlike carvel building, where planks are laid edge-to-edge, clinker construction utilizes overlapping planks, creating a remarkably strong and flexible hull. This overlapping, reminiscent of fish scales, gives the vessel its characteristic appearance and contributes to its seaworthiness. The process begins with meticulously shaped planks, each carefully fitted to its neighbor. This precise fitting requires exceptional skill and understanding of wood properties. Each plank is not merely placed; it's expertly shaped and positioned to contribute to the overall strength and watertightness of the vessel.

The construction itself is a fascinating dance of craftsmanship and precision. Experienced shipwrights, utilizing hand tools primarily, meticulously fasten each plank to the ribs or frames of the vessel. This fastening is traditionally done with rivets or wooden pegs, a testament to the ingenuity of the Viking builders. The strength of the clinker hull lies not only in the overlapping planks, but also in the way the planks are fastened and the overall design of the hull. The design accounts for the flex and stresses encountered at sea, allowing the vessel to handle rough waters with greater ease than many other designs of the time. This resilience was crucial for Viking longships navigating the unpredictable North Atlantic.

Furthermore, the clinker building technique offers several advantages beyond its strength and flexibility. The overlapping planks provide inherent watertightness, minimizing the need for extensive caulking. This is particularly important in harsh sea conditions where leaks could be devastating.

The method also allowed for easier repair at sea. A damaged plank could often be replaced relatively easily without needing to dismantle large sections of the hull—a practical advantage in the unpredictable and often unforgiving environments where Viking ships operated. The flexibility of the design, the overlapping planks absorbing impact rather than fracturing like a rigid carvel hull, further contributed to its resilience.

Beyond its practical advantages, the construction of a clinker-built vessel represented a significant investment of time, skill, and resources. It required a deep understanding of naval architecture and a mastery of woodworking techniques. The process would often involve a team of highly skilled craftsmen, each specializing in different aspects of construction. The design process itself was meticulous, the shaping of each plank to precise measurements a testament to the precision of Viking boat builders. The construction of a longships was not just a construction project; it was a collective effort demonstrating social cooperation and a deep connection to maritime traditions.

In essence, clinker building transcends mere craftsmanship; it represents a profound understanding of both material science and naval architecture. The overlapping planks create a hull remarkably resistant to damage from waves and impact. The method allows for efficient construction and easier repair, crucial factors influencing the success of Viking voyages. The design's adaptability to varying conditions allowed Viking longships to venture far from their home shores, contributing significantly to their expansion and influence across Europe and beyond. The elegance and resilience of clinker-built vessels serve as a testament to the skill and ingenuity of Viking shipwrights. Their legacy remains not just in the historical record, but in the enduring influence of their construction techniques on later boat-building traditions. The method speaks to a time of innovation and adaptation, where practicality and aesthetic grace found a harmonious balance. The clinker-built ship, therefore, stands as a powerful symbol of Viking ingenuity and maritime prowess. Its construction remains a subject of study and admiration, reflecting the mastery of design, engineering and craftsmanship of the Viking Age. The study of clinker building allows for a deeper understanding of the technological and cultural landscape of the Viking Age.

4 Chapter 4: Medieval Marvels

The medieval period, often associated with castles and knights, also witnessed significant advancements in shipbuilding. These weren't merely incremental improvements; they represented a fundamental shift in maritime capabilities, paving the way for the Age of Exploration. Larger, more robust vessels emerged, capable of carrying heavier loads and weathering harsher storms than their predecessors. This expansion of capabilities directly impacted trade, warfare, and exploration.

Consider the cog, a quintessential medieval ship. Its design, a departure from earlier, simpler vessels, incorporated a full-bodied hull, significant improvements in the structural strength through advanced techniques in planking, and a high sternpost which significantly improved stability, particularly in challenging sea conditions. This sturdy construction enabled longer voyages, opening up new trade routes and facilitating the movement of greater quantities of goods across vast distances. The cog's impact extended beyond commerce; its robust nature made it suitable for warfare as well, leading to its adoption for military purposes. The improved seaworthiness and larger cargo capacity transformed maritime trade. It wasn't just about transporting more goods; it meant that more diverse and valuable cargoes could be shipped reliably and safely across longer routes. This impacted the economies of nations and increased interconnectivity in Europe and beyond.

But the cog was not the only marvel of medieval shipbuilding. The carrack, a larger and more sophisticated vessel, represents a further step in this evolution. Characterized by its tall sterncastle – a fortified structure at the rear of the ship which allowed for better defense and navigation – and its three or more masts, the carrack offered improved maneuverability and an increased carrying capacity, compared even to the already impressive cog. This design offered superior speed, allowing for quicker voyages and increased trade opportunities. The carrack's impact was profound, fundamentally changing both trade and naval warfare. Its ability to transport larger quantities of goods across longer distances spurred economic growth and facilitated the exchange of ideas and cultural artifacts. On the battlefield, it became a powerful warship, capable of carrying significant numbers of soldiers and weaponry.

The construction techniques employed in building these vessels were as remarkable as the ships themselves. Shipwrights, highly skilled craftsmen, utilized advanced carpentry techniques, employing improved frames and planking methods for both the cog and carrack. These innovations ensured the strength and durability necessary for these larger, more ambitious vessels. This advanced shipbuilding was not an isolated endeavor, it depended on the work of multiple specialized artisans working in coordination to create the final product, a practice common across all eras of shipbuilding. Understanding this requires acknowledging the communal work and specialized tasks, moving from hull design, to rigging, and all other elements needed to make a seaworthy vessel.

Furthermore, medieval shipbuilding wasn't simply about bigger ships; it also involved improvements in rigging and navigational techniques. The development of more effective sails and rigging systems allowed for better control and maneuverability, while advancements in navigation, such as the improved use of the astrolabe and magnetic compass, extended the range of voyages. This advancement allowed sailors to navigate more precisely. These tools enabled more efficient and reliable travel, enabling access to new trade routes, and making transoceanic voyages more feasible than ever before. This allowed for the expansion of trade networks, which, in turn, facilitated the exchange of goods, ideas and cultures across Europe and beyond.

In conclusion, the medieval period was a pivotal time for shipbuilding. The cog and carrack represent not just larger vessels, but a significant leap in maritime technology and a testament to the ingenuity and craftsmanship of medieval shipwrights. These improvements, combined with advancements in navigation, laid the groundwork for the Age of Exploration and significantly impacted the course of history. The legacy of these medieval marvels continues to resonate, reminding us of the importance of innovation and human ingenuity in pushing the boundaries of exploration and trade. Their evolution showcases the constant drive to improve maritime technology, impacting every subsequent era of shipbuilding. The lasting impact of these medieval vessels underscores their crucial place in maritime history and their pivotal role in shaping the world we know today.

The legacy of these remarkable ships continues to inspire and fascinate, highlighting the enduring power of human ingenuity and innovation.

4.1 Cog Ships

The cog, a dominant vessel of the medieval period, represented a significant leap in shipbuilding technology. Its robust construction and impressive cargo capacity revolutionized maritime trade and warfare across Europe. Unlike earlier designs, the cog boasted a full-bodied hull, offering increased stability and the ability to withstand the rigors of long voyages and stormy seas. This substantial design allowed for the carriage of significantly larger quantities of goods, driving economic expansion and facilitating the growth of international commerce. The increased stability also made it a more effective warship, capable of supporting heavier armaments and withstanding the impacts of naval combat more effectively than its predecessors.

Its construction involved meticulous craftsmanship. Shipwrights, skilled artisans, employed advanced techniques to create a strong and durable vessel. The hull, typically constructed using oak planks, was reinforced with sturdy frames. These frames provided structural integrity, crucial for enduring the stresses of navigating unpredictable seas and carrying substantial loads. The quality of the timber was paramount; carefully selected, seasoned oak ensured the longevity and seaworthiness of the ship. This meticulous attention to detail demonstrates the high level of expertise within the medieval shipbuilding community. The creation of each cog was a collaborative effort, requiring a team of skilled individuals with specialized knowledge and abilities.

One key innovation was the use of a strengthened sternpost, allowing for more effective steering. Precisely fitted and secured, this component played a vital role in maneuverability and control, especially in challenging conditions. This improvement reduced the reliance on oars for steering, allowing for greater flexibility and speed, particularly when sailing under the power of wind. Efficient steering was crucial for navigating intricate waterways and for responding effectively to rapidly changing weather patterns and conditions at sea. Accurate steering was essential for commercial voyages, ensuring that vessels could reach their destinations promptly and efficiently.

Beyond its structural improvements, the cog also benefited from advanced rigging techniques. Sail plans were designed for optimum performance, maximizing efficiency and speed while maintaining stability. These improvements significantly reduced journey times and distances, further boosting maritime trade and fostering greater interconnectedness between various regions of Europe and beyond. The improvements in both design and construction helped ensure safer and more reliable voyages.

The impact of the cog on medieval society was profound. Its improved carrying capacity enabled larger-scale trade, fostering economic growth and cultural exchange. European economies prospered due to increased access to goods from other regions. The ability to transport large quantities of goods at lower cost fueled industrial development and stimulated innovation across multiple sectors. The increased mobility also fostered the growth and development of urban centers and the growth of international markets.

However, the cog was not without its limitations. Its size and design, while advantageous for cargo and stability, made it less maneuverable than smaller vessels. Its larger size and heavier construction presented a risk in shallow waters, making its range of operation in coastal waters and estuaries more limited. This presented a trade-off between the advantages of increased cargo capacity and the need for flexibility in navigation.

Despite its drawbacks, the cog undeniably marked a significant turning point in shipbuilding history. It played a crucial role in transforming European trade and warfare. This advancement showcased the continuous evolution of design and craftsmanship in ship construction and the enduring impact of innovation upon society. The cog's legacy lives on, not only in its historical importance, but also as a testament to the ingenuity and skill of medieval shipwrights. Its development paved the way for subsequent advancements, laying the foundation for the remarkable ships of later eras. The study of its construction offers valuable insights into the technological and economic forces that shaped medieval Europe. Its influence was far-reaching, a testament to its revolutionary design and construction. The cog remains a powerful symbol of medieval maritime prowess, reflecting the ingenuity and skill of its creators and the profound impact it had on the history of shipbuilding and global trade. The legacy of the cog continues to inspire awe and admiration for the artisans who brought this impressive vessel to life.

4.2 Carrack Design

The carrack, a majestic vessel of the late medieval period, represented a significant leap forward in shipbuilding technology. Its design, a blend of established techniques and innovative approaches, enabled voyages of unprecedented length and scale, paving the way for the Age of Exploration. The carrack's distinctive features, a product of centuries of maritime experience, weren't simply aesthetic choices; they were crucial for enhancing seaworthiness and cargo capacity. This wasn't just a matter of building a bigger boat; the carrack showcased a deeper understanding of naval architecture.

Unlike earlier vessels, the carrack boasted a significantly larger hull. This wasn't a mere increase in size, but a strategic design change. The longer hull, combined with a wider beam, provided greater stability, making it less susceptible to rolling and pitching in rough seas. This enhanced stability was crucial for carrying larger and heavier cargoes, dramatically increasing the profitability of long-distance trade. Improved stability also translated into safer conditions for the crew and passengers, an important consideration on journeys lasting months, even years. Furthermore, the increased cargo capacity facilitated the transportation of larger quantities of valuable goods, driving economic expansion across continents.

The carrack's sternpost rudder, a notable improvement over earlier steering systems, provided far greater control and maneuverability. This advanced steering mechanism, in combination with improved sail plans, allowed captains to navigate more effectively in diverse weather conditions. Accurate navigation, already crucial in earlier eras, became even more essential with the longer and more perilous voyages undertaken by carracks. They could now respond swiftly to changing winds and currents, ensuring the safety of the vessel and its valuable cargo. Consequently, the carrack opened up new trade routes and exploration possibilities previously deemed impossible with less maneuverable ships.

The carrack's three or four masts, carrying a substantial amount of sail area, gave it the capability to navigate reliably across vast oceans. Unlike their predecessors, these ships were equipped to handle strong headwinds, making voyages far more efficient. The arrangement of the masts and sails, a testament to the shipwrights' skill, was optimized for speed and handling. The design allowed for better control of the ship's speed and direction, essential in navigating unpredictable currents and weather patterns encountered on the open sea. This resulted in shorter journey times and reduced the likelihood of encountering adverse conditions for extended periods.

The construction of a carrack demanded considerable skill and craftsmanship. Shipbuilders painstakingly selected and prepared the timber, often using high-quality oak for the hull's frame. They utilized advanced techniques, like mortise and tenon joints, to create a strong and durable structure capable of withstanding the stresses of long voyages. This wasn't simply about building a robust frame; it involved a sophisticated understanding of material properties and structural mechanics. Every piece of timber, every joint, contributed to the overall integrity of the vessel, reflecting years of accumulated experience and expertise. The result was a ship capable of transporting significant cargo across treacherous oceans.

Beyond the purely structural aspects, the carrack's design incorporated features improving both safety and comfort for the crew. The inclusion of more substantial living quarters below deck and improved storage spaces, though seemingly minor details, markedly improved the quality of life for those aboard. Long voyages were notoriously harsh, but the enhanced conditions aboard carracks minimized hardship, improving crew morale and productivity, especially during periods of adverse weather conditions or extended voyages. This was especially important in attracting better-trained crew, contributing to safer and more successful voyages.

The development of the carrack epitomizes the constant evolution of shipbuilding techniques. It represents the culmination of centuries of maritime experience, blending practical innovations with sophisticated design principles. Its influence extended far beyond its immediate era. The carrack's design elements played a significant role in shaping the design of subsequent vessels, demonstrating its profound and lasting impact on naval architecture and the course of maritime history. Its legacy continues to inspire, reflecting a critical chapter in the history of maritime exploration and trade. These sturdy vessels were crucial instruments of global trade and exploration, transporting goods and people across oceans and fundamentally shaping the world we know today. Understanding its design is to understand a pivotal moment in human history.

5 Chapter 5: Age of Exploration

The caravel, a seemingly modest vessel compared to its later galleon counterparts, revolutionized exploration. Its design, a blend of lateen sails allowing for tacking into the wind and a sturdy hull capable of withstanding ocean swells, was the perfect synthesis of existing shipbuilding knowledge and innovative thinking. Smaller and more maneuverable than larger ships, caravels could navigate shallow coastal waters and explore intricate coastlines, opening up previously inaccessible regions. Portuguese shipwrights, particularly, perfected the caravel's design, making it a crucial element in their expansion across the Atlantic. Their meticulous attention to detail, gained through generations of maritime experience, ensured the reliability and seaworthiness essential for long voyages into the unknown. The caravel's success wasn't simply down to its design, however; it also required skilled crews and sophisticated navigational techniques.

This era witnessed a leap forward in navigation, driven by a desperate need for accurate charting during long voyages. The astrolabe, a tool used since antiquity, was refined for greater precision, enabling sailors to determine latitude through celestial observations. The development of improved charts, based on newly acquired geographical data, further reduced the risks of these perilous journeys. These advancements, coupled with the development of better cartography and the more accurate understanding of ocean currents and wind patterns, made the Age of Exploration possible. No longer were voyages limited by the fear of getting lost at sea; instead, they were calculated risks, based on a gradually increasing understanding of the world's oceans. The interplay between technological development and navigational expertise played a critical role in shaping the course of the age.

The Age of Exploration wasn't solely about discovery; it fundamentally reshaped global trade routes. Previously established trade networks were augmented, sometimes entirely superseded, by routes across the Atlantic and around the Cape of Good Hope. The flow of goods—spices, silks, precious metals—shifted dramatically, enriching European powers while profoundly impacting established trade centers in Asia and the Middle East. The caravel, and other similar vessels adapted for long voyages, became the vital arteries of these new trade networks, carrying not only valuable commodities but also people, ideas, and diseases. This globalization, though often exploitative, irrevocably altered the economic and political landscape of the world, a consequence directly linked to the advancements in shipbuilding and navigation that characterized this period. The scale of the transformation underscores the pivotal role of the ship in shaping history.

The impact of the Age of Exploration extended far beyond mere economics. European powers, emboldened by their improved shipbuilding techniques and navigational skills, established colonial empires across the globe. The consequences of this expansion were both profound and lasting, leaving an enduring legacy on the world's political systems, cultures, and demographics. This period of colonialism, fueled by the ability to travel across vast oceans, ushered in an era of unprecedented cultural exchange and the beginning of a truly globalized world. While this globalization was often deeply problematic and frequently oppressive, its impact on the interconnectedness of human societies is undeniable, making a direct link between the innovative shipbuilding of the era and the global transformations that followed. The ships weren't just tools; they were instruments of power, agents of change on a global scale.

The increased maritime traffic during this era also necessitated improvements in shipbuilding practices themselves. Shipyards grew in size and sophistication, often adapting techniques developed over centuries, fine-tuning them to meet the specific challenges of long-distance voyages. The construction of larger ships demanded more refined and sophisticated designs, focusing on enhancing both speed and seaworthiness. New techniques were developed, and old ones improved, reflecting the demand for ships capable of carrying greater cargo across longer distances. Shipbuilding became a strategic industry, directly linked to national power and wealth. This era saw a shift from localized, regional shipbuilding traditions to more standardized and efficient mass production techniques as demand increased for the construction of numerous vessels to meet the growing requirements of maritime trade and exploration. This shift laid the groundwork for the even larger industrialization of shipbuilding in later centuries.

5.1 Caravel Construction

The caravel, a seemingly simple ship, revolutionized 15th-century navigation. Its relatively small size, compared to later galleons, belied its remarkable capabilities. This was a vessel designed for speed, maneuverability, and surprisingly, long-distance voyages. Unlike the heavier, less agile cogs that preceded it, the caravel's design allowed it to navigate the unpredictable winds and currents of the Atlantic, opening up new worlds to European exploration. Its success rested on several key innovations.

The caravel's most distinctive feature was its lateen sail. This triangular sail, mounted on a long yard that could be adjusted to catch the wind from almost any direction, provided unparalleled control. Unlike the square sails of earlier vessels, which were effective only with favorable tailwinds, the lateen sail allowed caravels to tack against the wind, a crucial ability for oceanic voyages. This innovative design significantly reduced reliance on prevailing winds and allowed for much greater navigational flexibility, enabling explorers to chart new courses and reach distant shores previously considered inaccessible. Master shipwrights meticulously crafted these sails from durable canvas, ensuring their ability to withstand the relentless forces of the ocean.

The hull construction itself was equally crucial. Caravels typically employed a carvel-built hull, where planks were edge-joined and sealed, creating a smooth, watertight surface. This method, in contrast to clinker building, offered superior strength and reduced the ship's overall weight. The use of stronger hardwoods, carefully selected and seasoned, further enhanced the hull's durability, enabling the vessels to withstand the rigors of prolonged ocean voyages. The specific type of wood varied depending on availability and the resources of the shipwright; oak and pine were common choices but not universally applicable. These ships were also known for their relatively shallow draft, allowing them to navigate coastal waters and shallow inlets, advantages lacking in deeper-draft vessels.

Beyond the hull and sails, other design elements contributed to the caravel's effectiveness. The sternpost, a crucial structural element, was a point of significant refinement during this period. Its robust construction provided support for the rudder and enhanced the overall stability and maneuverability of the ship. The rudder's design, another key innovation, improved steering precision allowing for safer navigation through complex currents and coastal waters. A well-placed and robust rudder would make or break a journey on treacherous waters. In addition, the caravel's relatively lightweight construction, combined with its efficient sail plan, meant it required a smaller crew than larger ships of the time, resulting in considerable economic benefits.

But building a caravel wasn't a simple task. It required skilled artisans, working with precise tools and techniques handed down through generations of shipwrights. The process began with the selection of timber, which had to be of superior quality, free from defects and properly seasoned to prevent warping or cracking. The shipwright's understanding of the grain of the wood was paramount; the orientation of planks directly impacted the vessel's strength. Careful planning and precise measurements were crucial for the construction of the frame, which provided the skeleton of the ship. The ribs or frames were precisely cut and shaped, carefully fitted together to create a strong and durable structure.

Planking was a meticulous process, requiring the shipwright's expertise to ensure a watertight seal. Each plank was carefully shaped, fitted, and fastened to the frame. Caulking, the process of filling the gaps between the planks with oakum and pitch to prevent leaks, was another critical step. This required great care and skill to ensure a watertight hull, and any weakness here could lead to disaster. After the hull was completed, the deck was laid, the masts erected, and the rigging installed. The final stage of ship construction involved the fitting out, where provisions, equipment, and supplies were added. This process was crucial in ensuring crew safety and comfort during long voyages. The quality of workmanship significantly impacted the seaworthiness and longevity of the vessel. A poorly built caravel could prove disastrous, while a well-crafted vessel was a crucial tool in the Age of Exploration.

The construction of a caravel wasn't just about technical skill, it was a testament to the ingenuity and perseverance of the human spirit. It reflects a deep understanding of maritime principles and the ability to adapt and innovate within the constraints of available materials and technology. This vessel played a pivotal role in opening up new trade routes and expanding the horizons of exploration, making it a symbol of the human capacity for discovery. Its lasting influence on maritime history solidifies its place as one of the most important ship designs ever created. The legacy of the caravel speaks to the enduring power of human ingenuity and the critical role craftsmanship played in shaping the course of history.

5.2 Navigational Advancements

The Age of Exploration dramatically altered the course of history, fueled by advancements in navigation that allowed for unprecedented voyages of discovery. Before this era, sailors relied heavily on celestial navigation, using the stars and sun to estimate their position. However, this method was imprecise, susceptible to cloud cover, and often led to significant errors in reckoning. The development of more accurate instruments fundamentally changed this; voyages that were once fraught with peril became more predictable and reliable.

The invention of the astrolabe, a sophisticated tool for measuring the altitude of celestial bodies, marked a pivotal moment. Early versions were clumsy and required considerable skill to use effectively, but over time, refinements improved accuracy and ease of use. By measuring the altitude of the sun or stars, sailors could determine their latitude, a crucial piece of information for navigating across vast oceans. This allowed for better charting of coastal routes and even for venturing further from known landmasses.

Further progress in cartography—the science of mapmaking—played an equally crucial role. The creation of more accurate and detailed maps, utilizing newfound geographical knowledge and improved surveying techniques, aided navigation significantly. These maps, although initially imperfect, became increasingly reliable as exploration continued, building upon existing knowledge and incorporating new discoveries. This iterative process of mapping and exploration formed a positive feedback loop, allowing for more ambitious journeys.

The development of the mariner's compass, although not originating in this era, underwent significant improvement during the Age of Exploration. Early compasses were notoriously unreliable, prone to magnetic declination and inaccuracies. However, advancements in magnetic needle technology, combined with improved methods of mounting and calibration, resulted in a more dependable navigational aid. This instrument, combined with the astrolabe, provided sailors with a more complete picture of their location, allowing for more precise course correction and reducing the risk of getting lost.

Another significant navigational advancement was the improved understanding of currents and winds. Sailors began to meticulously document prevailing wind patterns and ocean currents, gaining valuable knowledge that dramatically impacted their ability to plan efficient voyages. This data, coupled with their improved ability to determine location, allowed them to predict their arrival times more accurately and to plan their routes more strategically, taking advantage of favorable winds and avoiding dangerous currents. This knowledge was often carefully guarded, treated as a trade secret passed down through families of sailors, contributing to the success and profitability of certain seafaring ventures.

The integration of these advancements fostered a synergistic effect; each improvement built upon others, creating a virtuous cycle of progress. Better maps enabled more accurate use of the astrolabe and compass, which in turn facilitated more accurate charting, and this refinement of navigational tools and techniques facilitated the exploration and exploitation of vast maritime resources. The result was a rapid expansion of global trade routes, a more interconnected world, and a new era of human history.

The voyages of discovery undertaken during this era, made possible by these navigational advancements, would dramatically reshape the political, economic, and cultural landscape of the globe. The exchange of goods, ideas, and people across continents, spurred by the improvements in navigation, would leave an enduring legacy that continues to resonate in our modern world. The ability to accurately navigate oceans—a seemingly simple concept—was the cornerstone of a period of unprecedented exploration and transformation. These advancements were not mere technical innovations; they were instrumental in shaping the globalized world we inhabit today.

The legacy of the Age of Exploration is complex and multifaceted, encompassing both positive and negative aspects. While the advancements in navigation facilitated unprecedented exploration and trade, it also led to the exploitation and colonization of many lands and peoples. Understanding the full impact of this era requires acknowledging both its positive contributions and its darker legacies. The ability to traverse vast oceans efficiently and reliably had profound consequences, transforming the geopolitical landscape and initiating a period of sustained global interaction that continues to shape the present day. The precise navigation techniques that made this possible laid the groundwork for modern maritime industries and, indirectly, for many of the technological advancements we enjoy today.

The story of navigational advancements during the Age of Exploration is not merely a narrative of technical progress; it is a human story of ambition, ingenuity, and the enduring quest for knowledge and exploration. It is a reminder of the power of innovation, highlighting how even seemingly small advancements can have far-reaching and transformative consequences. The improved methods of navigation fundamentally altered the course of history, creating a legacy that continues to influence our globalized society in countless ways. This ongoing influence is a testament to the profound impact of these navigational advancements, shaping the very fabric of our world. The seafaring prowess born from this period remains a source of wonder and inspiration, highlighting the human capacity for both exploration and transformation. The Age of Exploration stands as a powerful example of how innovation, driven by human curiosity, can lead to epoch-making changes across cultures and continents.

5.3 Global Trade Routes

The Age of Exploration, a period brimming with ambition and discovery, fundamentally reshaped global trade. Previously confined to regional networks, commerce now extended across oceans, connecting continents in unprecedented ways. This expansion was inextricably linked to advancements in shipbuilding, with the caravel playing a pivotal role. Its design, combining maneuverability and seaworthiness, allowed for exploration of previously unreachable coastlines. The Portuguese, masters of navigation and shipbuilding, utilized caravels extensively, establishing lucrative trade routes along the African coast and eventually reaching India. Their success spurred competition, leading other European nations to invest heavily in their own naval capabilities and maritime expansion.

The establishment of these new trade routes didn't just involve ships; it demanded a sophisticated infrastructure. Port cities flourished, transforming into bustling hubs of activity, supporting the needs of the burgeoning global trade. These ports became centers for ship repair, provisioning, and the exchange of goods, acting as crucial nodes in a vast network spanning the globe. The sheer scale of this activity required intricate logistical systems, demonstrating remarkable organizational capabilities. Efficient management was crucial for optimizing the flow of goods and ensuring profitability, leading to the development of complex trading companies. These companies, such as the Dutch East India Company, possessed immense power and influence. Their activities extended beyond mere trade; they engaged in colonization, establishing trading posts and settlements across continents. This colonization, often brutal and exploitative, was directly fueled by the economic opportunities presented by global trade routes. The wealth generated fuelled further investment in shipbuilding and exploration, creating a positive feedback loop that propelled the Age of Exploration forward. This complex interplay between trade, colonization, and technological advancement fundamentally reshaped the world's political and economic landscape.

However, the establishment of these routes was not without its challenges. Navigation remained a significant hurdle, particularly in the open ocean. The development of better navigational tools, such as the astrolabe and the improved mariner's compass, was crucial for safe and efficient voyages. Understanding weather patterns and ocean currents proved equally vital, requiring careful observation and the accumulation of practical knowledge passed down through generations of sailors. Moreover, the unpredictable nature of the sea itself posed constant threats, with storms and shipwrecks resulting in significant loss of life and cargo.

But the rewards far outweighed the risks. The influx of new goods – spices, silks, and precious metals – revolutionized European economies and societies. This influx fueled a period of unprecedented economic growth and expansion. The introduction of new crops and livestock, brought from far-flung corners of the globe, also transformed agricultural practices and contributed to population growth. These global trade routes facilitated the exchange of not only goods but also ideas, technologies, and cultural practices, ushering in an era of unprecedented intercultural exchange, albeit often under conditions of significant power imbalance. The impact of these global trade routes resonates even today, shaping patterns of globalization and international relations. The legacy of this transformative period continues to influence the modern world in ways both visible and subtle.

The sheer volume of goods being transported along these routes required larger and more capable ships. The caravel, while revolutionary, eventually gave way to larger vessels better suited for long voyages and carrying larger quantities of cargo. The carrack and the galleon, with their increased size and improved design, became prominent features of the global trade fleets. This evolution in shipbuilding technology further exemplifies the dynamic relationship between commerce and technological progress.

The drive for efficiency and profitability directly propelled advancements in naval architecture. Shipbuilders constantly sought ways to improve speed, cargo capacity, and seaworthiness, fostering innovation and driving improvements in construction techniques.

Finally, understanding these global trade routes requires acknowledging their impact on different societies. While European powers profited immensely, the consequences for indigenous populations were often devastating. Colonization, forced labor, and the disruption of traditional economies had a profound and often negative impact on the lives and cultures of many communities. Examining this history requires a critical perspective, acknowledging both the economic advancements and the human costs associated with the establishment of global trade routes. A comprehensive understanding demands a nuanced perspective, acknowledging the complexities and contradictions inherent in this pivotal period of global history. The long-lasting effects of this era continue to shape the world we live in today, reminding us of the profound impact of trade and technology on human societies.

6 Chapter 6: The Golden Age

The 16th and 17th centuries, often dubbed the Golden Age of Sail, witnessed a zenith in shipbuilding. Galleons, majestic vessels renowned for their size, power, and elegance, dominated the seas. These weren't merely improvements on earlier designs; they represented a significant leap forward in naval architecture and construction, reflecting a confluence of technological advancements and evolving maritime needs. Their construction, a testament to human ingenuity and the mastery of complex woodworking techniques, showcases a pivotal moment in maritime history.

Their sheer size was impressive, often exceeding 100 feet in length and displacing hundreds of tons of water. This scale demanded sophisticated shipbuilding methods, necessitating highly organized shipyards and teams of skilled artisans, many working within a highly structured and competitive framework. The increased size also necessitated improved hull designs for greater strength and seaworthiness. The development of robust framing techniques and the incorporation of stronger timbers allowed for the creation of larger, more resilient hulls capable of withstanding the rigors of long ocean voyages and fierce naval battles.

One of the key features of galleon construction was the use of more durable, stronger hardwoods and timbers than were previously standard. These were carefully selected and shaped, often employing advanced techniques of timber bending to create curved components like ribs and frames. The process involved specialized tools and expertise passed down through generations of shipwrights. Precise measurements and meticulous joinery were essential, ensuring a strong, watertight hull. The sheer volume of material needed for a single ship is astounding, highlighting the scale of operation and the significant economic impact of shipbuilding at this time.

Beyond the hull, the rigging of a galleon presented further engineering challenges. With multiple masts and an elaborate network of sails, galleons required highly skilled riggers who understood the dynamics of wind and sail management. The design of the rigging system itself evolved during this era. The improvements in the number and design of sails, along with the use of more advanced ropes and blocks, resulted in improved maneuverability and speed. This technological refinement contributed heavily to the dominance of European navies during this period.

Beyond sheer size and speed, the galleon's design often incorporated weaponry, reflecting its critical role in naval warfare. Gunports, carefully integrated into the hull design, provided firing positions for cannons, and the placement and design of these gunports were crucial for maximizing firing arcs and minimizing damage to the hull from recoil. The increasing size and power of cannons, along with the development of new ammunition and firing techniques, significantly impacted the design and construction of warships, creating a perpetual cycle of technological advancement in response to the evolving demands of naval combat. Therefore the galleon wasn't merely a vessel for transport, but a complex interplay of naval architecture, engineering, and military strategy.

The growth of dedicated shipyards reflected the increasing demand for galleons. These weren't simply places where ships were built; they were sophisticated industrial complexes that utilized specialized tools, machinery, and divisions of labor. The creation of these dedicated centers for shipbuilding facilitated efficiency, economies of scale, and also created a centralized training ground for shipwrights. Their development, moreover, highlights the increasing importance of maritime trade and naval power and the shift towards larger scale industrial production. This, in turn, fuelled competition and further technological innovation, ultimately leading to the production of ever-more advanced vessels.

The Golden Age of Sail wasn't just about the galleon; it also marked significant advancements in other shipbuilding technologies. Improvements in navigational instruments, like the astrolabe and the quadrant, enabled longer and more accurate voyages. Advancements in cartography, too, played an essential part in shaping the era, improving sailors' ability to chart courses and navigate safely across vast oceans. The combination of these factors propelled European maritime exploration and expansion across the globe. The era's shipbuilding innovations thus impacted not only naval power and trade, but also the course of global history, leading to unprecedented levels of interconnectivity and exchange.

In summary, the Golden Age of Sail represents a remarkable period in the history of shipbuilding. The development of the galleon, along with parallel innovations in naval architecture, materials, and construction techniques, transformed maritime warfare and international trade. It marked a high point in the craftsmanship of wooden shipbuilding, leaving behind a legacy that continues to inspire and fascinate even today. The sheer scale and complexity of the vessels built during this era demonstrate the power of human ingenuity and the crucial role of technology and craftsmanship in shaping the course of history.

6.1 Galleon Building

Galleons, the majestic symbols of the Spanish Armada and the wider Age of Sail, weren't simply built; they were meticulously crafted. Their construction represented the pinnacle of wooden shipbuilding, a testament to centuries of accumulated knowledge and skill. Consider the sheer scale: these were massive vessels, often exceeding 100 feet in length, demanding immense resources and a highly organized workforce. The process began long before the first plank was laid, with the selection of the finest timber, a critical factor influencing the ship's strength, durability, and longevity. Oak, renowned for its strength and resistance to rot, was the preferred choice for the keel, ribs, and planking. Other hardwoods, like pine and elm, were used strategically for masts, spars, and decking, carefully chosen for specific properties suited to each component.

The construction itself was a complex ballet of precision and craftsmanship. First came the keel laying, a ceremonial act marking the start of construction, after which the master shipwright and his skilled team worked in harmony, carefully shaping and assembling the ribs, forming the backbone of the vessel. These were not simply straight pieces of wood but carefully curved frames, meticulously shaped to the precise contours of the design, reflecting years of accumulated knowledge and experience. Their alignment was crucial to the overall structural integrity of the ship, with any deviation potentially leading to disastrous consequences at sea. The ribs were then connected to the keel, creating the skeleton that would support the entire structure.

The planking, a labor-intensive process, followed next. Each plank, carefully hewn and shaped, was meticulously fitted to its neighbors, creating a watertight hull. This wasn't simply a matter of nailing boards together. Shipwrights used advanced techniques such as caulking, packing fibers between the planks to prevent leaks, ensuring the vessel's seaworthiness. The precision required was remarkable; a single poorly fitted plank could compromise the entire hull's integrity. The sheer number of planks in a galleon, often thousands, was a testament to the scale of the endeavor and the level of skill needed.

Beyond the basic structure, the galleon's construction involved intricate details that reflected both aesthetic concerns and practical considerations. The deck, usually two or three levels, provided space for crew quarters, stores, and weaponry. The masts, towering tall above the deck, supported an extensive system of sails, harnessing the power of the wind to propel the ship across vast oceans. The rigging, a complex network of ropes and sails, required expertise to efficiently handle the tremendous forces involved. The sheer complexity of rigging a galleon, with its intricate network of ropes and pulleys, highlights the highly skilled labor that went into its creation. Consider the sheer weight of the masts and sails; the system needed to be robust yet agile enough to respond to the changing winds. Master riggers, specialists in their craft, were essential to the building process. Finally, the galleon was equipped with cannons, a formidable array of weaponry that solidified its status as a warship. The placement and integration of these weapons were carefully planned and executed to maximize their effectiveness, emphasizing the practical necessities alongside the aesthetic considerations of ship design. The entire process, from keel laying to the final touches on the weaponry, often lasted years, often involving multiple skilled trades, underlining the collaborative nature of galleon construction. The construction of a galleon wasn't merely shipbuilding; it was a complex logistical and engineering feat, demanding specialized skills, meticulous planning, and the coordinated efforts of hundreds of individuals. The final product, a majestic galleon, represented the epitome of maritime technology and engineering. It was a testament to human ingenuity and the power of collaboration, and a crucial component in maritime history's unfolding.

6.2 Shipyard Innovations

The seventeenth and eighteenth centuries witnessed a surge in shipbuilding innovation, driven by escalating naval competition and the expanding global trade. Shipbuilders weren't simply replicating existing designs; they were actively experimenting, pushing boundaries, and refining techniques honed over centuries. This era saw the refinement of existing designs, like the galleon, making them even more powerful and seaworthy. Increased size, improved maneuverability, and greater firepower were constantly sought. The demand for faster, more efficient ships led to significant advancements in hull design and rigging. These improvements were not merely aesthetic—they directly impacted a vessel's speed, stability, and ability to withstand the rigors of long voyages. This period also saw increased use of specialized tools and techniques, improving the accuracy and speed of construction, leading to larger and better-equipped ships.

One notable innovation was the development of more sophisticated rigging techniques. Sails became larger and more efficient, allowing ships to harness the wind more effectively, while advanced rigging systems enabled better control and maneuverability. Experimentation with different sail shapes and arrangements led to significant increases in speed and efficiency. These improvements were not made in isolation; shipwrights collaborated, sharing knowledge and techniques across shipyards and nations. The rise of naval academies and specialized training programs further facilitated this exchange of information, fostering a period of rapid technological advancement. This period was marked by the rise of standardized plans and designs, which facilitated mass production in shipyards that could produce numerous ships of a uniform quality and design.

The introduction of new materials also played a crucial role. While wood remained the primary building material, experimentation with different types of wood, metal fastenings, and improved caulking techniques enhanced durability and longevity. The gradual adoption of iron fastenings, for instance, while initially slow, eventually led to stronger, more reliable ships less prone to damage. The careful selection of timber, ensuring the best quality for each part of the ship, combined with advanced joinery techniques, enhanced the structural integrity of vessels. This careful consideration of materials highlighted the evolving understanding of materials science in shipbuilding. The use of different types of wood, such as oak for frames and pine for planking, reflected a growing awareness of the specific properties of various materials and their suitability for particular applications within the ship's structure. The focus on strength and durability enabled the construction of larger vessels that could carry heavier cargo and endure more arduous voyages.

Beyond the materials themselves, the construction techniques underwent significant transformations. The introduction of more precise tools and measuring instruments, like improved saws, chisels, and adzes, allowed for more accurate and consistent construction. This, in turn, improved the overall quality of the vessels produced. These advancements were not solely about tools but also about the workforce. The rise of specialized labor within shipyards saw the emergence of skilled craftsmen, each responsible for specific tasks, leading to a more efficient and effective division of labor. Increased organization within shipyards led to greater productivity and the ability to build larger and more complex ships in a shorter time frame. The precise assembly of components, from the keel laying to the final fitting out, showcased the refined skills and collaboration within the shipbuilding community.

The growth of shipyards themselves mirrored these advancements. Larger, better-equipped yards emerged, capable of handling larger vessels and employing more skilled labor. The expansion of shipyards and associated industries led to the development of specialized ports and infrastructure, facilitating greater shipbuilding output and supporting the burgeoning maritime trade. These improvements were not confined to a single nation. The exchange of ideas and techniques across international boundaries led to a period of unprecedented progress in shipbuilding technology. This cross-cultural exchange fostered competition and innovation, leading to a rapid pace of advancements across the global shipbuilding community. The growing sophistication of shipbuilding practices during this period reflected not only technological advancements but also a broader understanding of maritime engineering, demonstrating a significant shift in how ships were designed, constructed, and utilized. This marked the beginning of a more scientific approach to shipbuilding, laying the groundwork for the further technological advancements that would characterize the centuries to come.

7 Chapter 7: Modern Methods

Steel replaced wood as the primary shipbuilding material during the 20th century, revolutionizing the industry. The shift wasn't merely a change in material; it represented a fundamental alteration in construction methods, requiring new skills and technologies. Instead of skilled artisans shaping timber, engineers now employed advanced calculations and sophisticated machinery. This transformation led to larger, stronger, and more efficient vessels, capable of carrying heavier loads and traversing greater distances than their wooden predecessors. The introduction of steel allowed for the development of standardized designs, contributing to the rise of mass production in shipyards. This efficiency significantly reduced construction time and costs, impacting global trade and transportation.

The advent of powerful welding techniques further cemented steel's dominance. Riveting, the traditional method of joining metal plates, was gradually replaced by welding, providing stronger and more watertight joints. These welds, carefully executed, eliminated many potential points of leakage and structural weakness. Furthermore, the introduction of automated welding systems boosted production speed and consistency, making it possible to build larger and more complex ships with greater accuracy. The transition to steel also necessitated new safety protocols and worker training, as welding involves significant risks, unlike the comparatively less dangerous woodworking of previous eras.

Modern shipyards stand as testaments to technological advancements. No longer are they small, scattered operations; instead, they have evolved into highly specialized industrial complexes. These enormous facilities utilize sophisticated machinery, from massive cranes capable of lifting hundreds of tons to precisely calibrated robotic welders. The integration of computer-aided design (CAD) and computer-aided manufacturing (CAM) has streamlined the design and construction process. Sophisticated software packages allow engineers to model entire ships virtually, simulating stress, strain, and other critical parameters before physical construction begins. This virtual prototyping significantly reduces design errors and improves efficiency. The implementation of these advanced technologies has necessitated a skilled workforce capable of operating and maintaining this complex machinery. This requires a shift in the workforce from traditional carpentry skills to a strong base in engineering and computer-aided design.

Beyond steel, the use of advanced materials like composite materials—fiber-reinforced polymers, for instance—is gaining momentum. These materials offer a compelling combination of high strength-to-weight ratios, corrosion resistance, and durability. Their use allows for the creation of lighter, faster, and more fuel-efficient vessels, essential in a world increasingly focused on sustainability. Moreover, the ability to tailor composite materials to specific applications opens up new design possibilities. The use of advanced composites is particularly prevalent in the construction of smaller vessels, high-speed crafts, and specialized applications where weight is a critical consideration. However, the higher cost of these materials and the complexities of manufacturing processes pose challenges to their widespread adoption.

The future of shipbuilding is likely to be defined by automation, sustainability, and the integration of new technologies. Autonomous vessels, capable of navigating and operating without human crews, are rapidly progressing from concept to reality. This automation has the potential to revolutionize shipping, reducing costs, increasing efficiency, and even improving safety in dangerous maritime conditions. Likewise, there is a growing emphasis on sustainable practices. The industry is focusing on reducing its environmental impact through the development of more fuel-efficient designs and the use of eco-friendly materials and energy sources. Innovations in hull design, propulsion systems, and waste management are crucial in achieving these sustainability goals. Ultimately, the future of shipbuilding hinges on addressing global challenges, including those related to climate change, resource management, and technological innovation. Adapting to these challenges will be crucial for ensuring the continued growth and success of this vital industry.

7.1 Steel Shipbuilding

The shift from wood to steel marked a pivotal moment in shipbuilding history, a revolution driven by the inherent strengths of steel and the burgeoning industrial revolution. Steel's superior strength-to-weight ratio allowed for the construction of larger, faster, and more robust vessels, fundamentally altering maritime capabilities and global trade. The transition wasn't immediate; early steel ships often incorporated wooden elements, a testament to the ingrained expertise and familiarity with wooden construction methods. However, the advantages of steel were undeniable, and its adoption accelerated rapidly.

Early steel shipbuilding techniques involved riveting, a process of joining individual steel plates with overlapping edges fastened by rivets. This method, though labor-intensive, proved incredibly strong and reliable, leading to the development of standardized shipbuilding practices and the establishment of large-scale steel shipyards. These yards, with their specialized tools and processes, represented a significant technological advancement, paving the way for mass production and economies of scale previously unimaginable in the wood-based shipbuilding tradition. The resulting increase in ship production dramatically altered maritime commerce and the geopolitical landscape.

The development of steel shipbuilding was intricately linked to advances in metallurgical engineering. The production of higher-quality, more consistent steel was crucial for the success of steel ships. Innovations in steelmaking processes, like the Bessemer process, dramatically reduced the cost and improved the quality of steel, making it a viable alternative to wood. This availability of high-quality steel, coupled with advancements in structural engineering principles, allowed naval architects to design more sophisticated ship hulls, optimized for strength, speed, and cargo capacity. It was a symbiotic relationship, with progress in metallurgy di-

rectly translating to more efficient and effective ship designs.
Beyond the material itself, the transition to steel necessitated the development of new tools and techniques. Specialized riveting machines, for instance, significantly increased the speed and efficiency of hull construction. The rise of steel shipbuilding also led to innovations in design software and fabrication methods. Computer-aided design (CAD) and numerical control (NC) machining, developed later, revolutionized the precision and speed of ship construction. The use of prefabricated components and standardized parts allowed for faster assembly and reduced overall construction time, lowering costs and improving production efficiency.
Steel's impact extended far beyond mere material substitution. Its adoption facilitated the development of larger and more complex vessel designs previously impossible with wood. This led to a significant increase in cargo capacity and range, transforming global trade and drastically accelerating the process of globalization. The construction of enormous ocean liners and massive cargo ships became a reality, fundamentally altering the logistics of international commerce and significantly impacting global economies.
The introduction of steel also revolutionized naval warfare. Steel warships, with their superior strength and firepower, quickly surpassed wooden vessels in terms of effectiveness. The construction of powerful battleships and cruisers, made possible by the strength and durability of steel, became a defining feature of 20th-century naval power. The impact on naval strategy and tactics was profound; the ability to construct more robust and heavily armed vessels fundamentally altered the nature of naval engage-

ments.

However, the transition to steel shipbuilding wasn't without its challenges. The initial learning curve was steep, and adapting established shipbuilding practices to steel presented significant obstacles. Working with steel demanded new skills and expertise, requiring extensive training for shipwrights and other skilled laborers. Furthermore, early steel ships encountered problems related to corrosion, which demanded the development of effective protective coatings and maintenance procedures.

Despite these challenges, the advantages of steel shipbuilding proved insurmountable. The increased strength, durability, and efficiency of steel ships revolutionized the maritime industry. It allowed for the construction of larger, faster, and more efficient vessels, fundamentally transforming global trade, naval warfare, and the overall maritime landscape. The evolution from simple rivets to sophisticated welding techniques and the integration of advanced materials mark a continuing story of innovation and refinement in this crucial industry. And the story continues to evolve even today, with ongoing advancements in steel technology and shipbuilding techniques promising even more significant advancements in the future of maritime transport. The enduring legacy of steel shipbuilding is evident in the ships that still ply the world's oceans, a testament to its enduring impact on history and technology.

7.2 Engineered Vessels

The transition from wooden hulls to steel marked a paradigm shift in shipbuilding. No longer constrained by the limitations of timber, naval architects and engineers could design vessels of unprecedented size and strength. The introduction of steel allowed for the creation of larger cargo ships, capable of carrying vastly increased quantities of goods across oceans, fundamentally altering global trade patterns and significantly impacting global economies. Furthermore, the development of sophisticated welding techniques enabled the construction of more complex and efficient hull forms. This allowed for faster speeds, improved maneuverability, and enhanced structural integrity. The resulting vessels could withstand harsher weather conditions and navigate challenging waterways with greater safety, leading to a dramatic expansion of maritime commerce and exploration. The use of steel also allowed for the integration of powerful engines, further enhancing efficiency and range. This in turn facilitated the growth of the global shipping industry.

This era also witnessed the rise of sophisticated computer-aided design (CAD) software. These programs allowed engineers to model and simulate vessel performance before construction, significantly reducing design flaws and errors. This precision engineering yielded lighter, stronger, and more fuel-efficient ships, contributing to environmental sustainability and economic viability. Moreover, CAD systems enabled the efficient sharing of design information amongst teams, streamlining the collaborative process and accelerating the construction timeline. This accelerated project completion and the resulting increased productivity benefitted the shipbuilding industry greatly. This ability to precisely design components, to optimize structural integrity, and to simulate real-world conditions significantly reduced production costs and materials waste. This new precision engineering meant that fewer materials were used and less labor was wasted.

The development of new materials beyond steel further enhanced the possibilities of engineered vessels. Aluminum alloys, for instance, offered a lighter weight alternative while maintaining significant strength. This characteristic made aluminum particularly suitable for high-speed craft and vessels requiring reduced fuel consumption. Advanced composites, combining the properties of fibers (like carbon fiber or fiberglass) with resin matrices, provided another option. These materials offer impressive strength-to-weight ratios, enabling the construction of lighter, faster, and more durable vessels. Their corrosion resistance also extends the lifespan of the vessel, minimizing maintenance requirements. Consequently, these newer materials, like aluminum and composites, represent advancements in naval architecture, allowing for vessels that are both more efficient and more sustainable.

Modern shipyards themselves underwent a technological transformation. Automated processes, robotic welding, and computer-controlled cutting machines increased both the speed and precision of construction. These advances significantly reduced human error, while also boosting productivity. Moreover, this technology contributed to improved worker safety in what was previously a dangerous occupation. The integration of such advanced technology in shipyards has not only drastically improved efficiency but also has resulted in higher-quality vessels that perform more reliably. Automated quality control measures ensure higher production standards, consistently improving the vessels being built.

The ongoing evolution of shipbuilding technology continues to shape the future. Research into new materials, such as advanced high-strength steels and bio-inspired designs, promises even more efficient and sustainable vessels. The integration of smart sensors and automated systems enhances vessel operation and maintenance. This leads to improved safety, reduced operating costs, and optimized performance. Furthermore, these developments contribute towards meeting emerging environmental regulations, promoting a more sustainable maritime industry. The future of shipbuilding promises even more sophisticated vessels adapted to the changing climate and the increasing demands of a globally connected world. The pursuit of ever-more efficient and robust designs pushes the boundaries of engineering, creating new possibilities for exploring, transporting, and utilizing the world's oceans.

7.3 Modern Shipyards

Modern shipyards represent a dramatic departure from their historical predecessors. Gone are the days of hand-hewn timbers and painstaking craftsmanship visible to the naked eye, replaced by a complex interplay of engineering, technology, and industrial scale. Today's shipyards are sprawling complexes, often employing thousands of workers and incorporating the latest advancements in design software, automated welding systems, and precision manufacturing. The sheer size and complexity of modern vessels, from colossal container ships to sophisticated cruise liners, necessitates this level of sophistication. Think about the sheer volume of steel used in construction, a testament to the industry's evolution from wooden hulls to modern giants.

The shift to steel construction revolutionized shipbuilding. No longer were builders limited by the size and strength of wood. Steel's strength-to-weight ratio allowed for the creation of far larger and more resilient vessels. This transition demanded new skills and technologies. Specialized welders, skilled in techniques like submerged arc welding, became essential, ensuring the integrity of massive steel structures. The process became significantly more precise, minimizing human error and maximizing efficiency. Advanced computer-aided design (CAD) software plays a crucial role in modern shipyard operations, allowing engineers to design complex hull shapes, optimize weight distribution, and simulate vessel performance before a single weld is made. This pre-fabrication strategy drastically reduces construction time and waste.

Prefabrication is a cornerstone of modern shipyard efficiency. Large sections of the hull and superstructure are constructed off-site in modular units, then transported and assembled in the dry dock. This assembly-line approach, mirroring advancements in other manufacturing industries, is key to the efficient and rapid construction of modern ships. It allows for parallel tasks, optimizing the use of resources and labor. This modular approach also allows for greater specialization within the shipyard, with specific teams focusing on specific components, leading to improved quality control and increased expertise in each stage of the construction. It is a far cry from the small, family-run shipyards of centuries past.

Furthermore, modern shipyards leverage sophisticated computer numerical control (CNC) machinery. These machines perform tasks with unparalleled precision, cutting and shaping steel components to exact specifications. This ensures consistency in construction and eliminates many of the inaccuracies that were inevitable in traditional methods. CNC technology allows for complex curves and intricate designs that would be impossible to replicate manually. The precision of CNC machining significantly reduces material waste and enhances the overall structural integrity of the vessel. The efficiency gains are substantial.

Beyond steel, modern shipyards are exploring advanced materials like high-strength aluminum alloys and fiber-reinforced composites. These materials offer benefits in terms of weight reduction, corrosion resistance, and increased durability. Aluminum alloys are particularly suitable for high-speed vessels, while composites are becoming increasingly common in specialized applications where weight is a critical factor. These choices often depend on the intended use of the vessel. The research and development invested in these materials are leading to even more advanced and efficient shipbuilding practices. The ongoing exploration of innovative materials promises a future of lighter, stronger, and more sustainable ships.

Automation is playing an increasingly significant role. While human expertise remains essential, robotics and automated systems are handling more and more tasks, from welding and painting to material handling and inspection. This not only increases efficiency but also enhances safety within the shipyard environment. Automated systems can perform repetitive or hazardous tasks with greater consistency and accuracy than human workers, leading to a safer and more productive work environment. However, the integration of these technologies requires careful planning and training of personnel.

Modern shipyards aren't merely construction sites; they are highly integrated manufacturing facilities. They rely heavily on sophisticated supply chain management systems, ensuring the timely delivery of materials and components. This complex logistical network, linking suppliers and subcontractors across the globe, is essential for the timely completion of projects. Any disruption to this system can have significant impacts on the shipyard's operations, highlighting the importance of robust logistics.

Finally, sustainability is becoming a paramount concern. Modern shipyards are exploring more environmentally friendly construction practices, reducing waste and emissions. The industry is moving towards more sustainable materials and energy-efficient technologies. Stricter environmental regulations and growing consumer awareness are driving this shift, placing a premium on eco-friendly shipbuilding methods. This is an area where innovation will increasingly play a crucial role in the future of the industry. Future designs will need to factor in environmental impact alongside the technical aspects of ship construction. The changes are significant and far-reaching. The future of shipyards is a blend of sophisticated technology, environmental awareness, and a continued focus on the craftsmanship which has defined the industry for centuries.

7.4 Advanced Materials

The transition to steel marked a significant leap in shipbuilding, offering unprecedented strength and durability. However, the search for even better materials continues, driven by the need for lighter, stronger, and more corrosion-resistant vessels. Modern ship construction now incorporates a wide array of advanced materials, each selected to optimize specific aspects of the vessel's performance. These materials are not simply replacements for steel; they represent a sophisticated integration of properties, allowing for increased efficiency and novel designs previously unimaginable.

Fiber-reinforced polymers (FRP), commonly known as composites, have become increasingly prevalent. Their exceptional strength-to-weight ratio makes them ideal for high-speed craft, where minimizing weight is crucial for fuel efficiency. Furthermore, their resistance to corrosion eliminates the need for extensive maintenance associated with steel hulls. Different types of FRP composites, using various fibers like carbon fiber, fiberglass, and aramid fibers, along with different resin systems, allow for tailored material properties to meet specific design needs. This flexibility allows shipbuilders to create custom composites, maximizing performance and reducing overall weight. For instance, a fast patrol boat might benefit from carbon fiber's high stiffness and strength, resulting in enhanced speed and maneuverability.

Aluminum alloys, another lightweight option, offer excellent corrosion resistance and high strength. They are particularly useful in applications demanding high strength and low weight, such as high-speed ferries or smaller naval vessels. Their ductility also allows for more complex hull designs and easier fabrication, leading to cost savings in construction. However, it's important to note that aluminum's susceptibility to fatigue under cyclical stress must be considered during the design phase, necessitating specialized welding and fabrication techniques. Proper design and quality control are paramount to prevent premature failure and ensure the longevity of the aluminum structure.

The use of high-strength steels, while still predominantly employed, has seen significant advancements. Modern high-strength steel grades exhibit superior tensile strength and yield strength compared to traditional mild steel, allowing for thinner plate thicknesses and reduced weight without sacrificing structural integrity. This translates to lower material costs and improved fuel efficiency. Moreover, research into advanced steel alloys is constantly improving corrosion resistance and fatigue performance, pushing the boundaries of what is possible with traditional steel construction.

Beyond the main structural materials, advanced materials also play crucial roles in other ship components. Ceramics are increasingly used in high-wear applications, such as propeller shafts and engine components, offering superior abrasion resistance and high-temperature tolerance. This reduces wear and tear, leading to extended component lifespans and reduced maintenance needs. Moreover, advanced coatings and surface treatments are employed to enhance the corrosion resistance and longevity of both steel and aluminum components, providing additional protection against the harsh marine environment.

The ongoing quest for improved material technologies in shipbuilding is driven by several factors. Fuel efficiency remains a paramount concern, with the marine industry striving to reduce its environmental impact. Lighter materials contribute directly to reduced fuel consumption, reducing operating costs and lessening the carbon footprint. Improved corrosion resistance also plays a vital role in reducing maintenance costs and extending the operational lifespan of ships. A ship that requires less frequent maintenance saves money and reduces downtime, improving the overall return on investment.

Moreover, the development of advanced materials allows for innovative design solutions. The flexibility and strength of composites enable the construction of more complex and streamlined hull forms, which can lead to improved hydrodynamic performance and reduced drag. This, in turn, further enhances fuel efficiency and speed. The ongoing research and development in the field of materials science continue to push the boundaries of what is possible, offering new possibilities for more efficient, durable, and sustainable ships. The future of shipbuilding relies heavily on the continued development and incorporation of advanced materials, allowing for the construction of vessels better suited to the demands of the 21st century and beyond. The integration of these innovative materials is not merely an advancement; it is essential for the continued growth and sustainability of the maritime industry. By thoughtfully considering and expertly applying these diverse materials, shipbuilders can create safer, more efficient, and environmentally responsible vessels.

7.5 Future of Shipbuilding

The shift towards automation is undeniable. Robotics and AI-driven systems are already transforming various aspects of shipbuilding, from automated welding and cutting to the precise placement of components. This isn't just about speed and efficiency; it's about minimizing human error and creating a safer work environment. Imagine a future where complex hull sections are assembled with laser-guided precision, ensuring a flawless, structurally sound vessel every time. This advanced automation promises greater consistency, reducing the need for extensive rework and minimizing material waste. The resulting cost savings could revolutionize the industry, making shipbuilding more accessible and cost-effective. Furthermore, the integration of digital twins—virtual representations of the ship—allows for thorough testing and simulation before a single rivet is driven, leading to even greater efficiency and optimized designs.

However, the human element remains crucial. While automation takes over repetitive tasks, skilled shipwrights and engineers are still indispensable. Their expertise in problem-solving, critical thinking, and adaptation to unforeseen challenges will always be valuable. The future shipyard will be a collaborative space, blending human ingenuity with machine precision. Think of it as an orchestra; the conductor (the human expertise) directs the automated instruments (robots and AI) to create a harmonious whole, a vessel of exceptional quality and efficiency. This integration will require a shift in education and training, focusing on the skills needed to manage and maintain these advanced systems while retaining traditional craftsmanship. We will need a new generation of shipbuilders who are proficient in both traditional techniques and advanced technologies.

Sustainable practices are no longer optional; they're imperative. The future of shipbuilding lies in environmentally friendly materials and processes. This translates to exploring bio-based composites, lightweight materials reducing fuel consumption, and developing more efficient propulsion systems. We are seeing advancements in the use of recycled materials and implementing closed-loop systems to minimize waste. The ships of tomorrow must be designed with a lower carbon footprint, reducing the industry's overall environmental impact. This will require innovation in both material science and ship design—a move towards greener technologies is not merely a trend, but an ethical necessity and a factor that will drive economic competitiveness. This shift may necessitate investment in research and development for new materials, and governments may play an essential role in incentivizing these developments through grants, tax breaks, or other policy initiatives.

The design of future vessels will be greatly influenced by artificial intelligence. AI algorithms can analyze vast datasets of ocean conditions, weather patterns, and operational requirements to create highly optimized designs. This means ships that are more fuel-efficient, robust, and better suited to their intended purpose. The optimization process can also incorporate considerations such as crew safety, environmental impact, and maintenance requirements. This data-driven approach, combined with advanced modeling techniques, allows for greater precision in forecasting and simulating various scenarios, enhancing safety and operational efficiency. The implementation of advanced analytics can also lead to improved predictive maintenance strategies, saving time and costs associated with unplanned downtime.

The geopolitical landscape will undoubtedly shape the future of shipbuilding. Growing global demand, particularly in developing nations, presents immense opportunities. However, this also presents challenges in terms of resource allocation, trade regulations, and geopolitical stability. Furthermore, the development of new technologies and materials will drive a competitive environment, necessitating continuous innovation and adaptation. This means focusing on research and development, investment in skilled labor, and collaboration across international boundaries. Stronger partnerships between nations could be instrumental in sharing knowledge, resources, and expertise to foster a more sustainable and globally equitable shipbuilding industry. Strategic alliances may prove more critical than ever in navigating complex political dynamics and fostering cooperation.

www.ingramcontent.com/pod-product-compliance
Lightning Source LLC
LaVergne TN
LVHW022015200926
848481LV00007B/2394
9798348595500